DESIGNING
BUILDING & TESTING
YOUR OWN
SPEAKER SYSTEM
WITH PROJECTS
2ND EDITION

Other TAB Books by the Author

DESIGNING
BUILDING & TESTING
YOUR OWN
SPEAKER SYSTEM
WITH PROJECTS
2ND EDITION

BY DAVID B. WEEMS

TAB BOOKS Inc.

BLUE RIDGE SUMMIT, PA. 17214

SECOND EDITION

FIRST PRINTING

Copyright © 1984 by TAB BOOKS Inc.

Printed in the United States of America

Library of Congress Cataloging in Publication Data

Weems, David B.
      Designing, building & testing your own speaker systems—
with projects.

      Includes index.
      1. Loud-speakers—Design and construction—Amateurs'
manuals. I. Title.
TK9968.W44   1984   621.38'028'2   84-8889
ISBN 0-8306-1964-X (pbk.)
ISBN 0-8306-0964-4 (Radio Shack ed.)

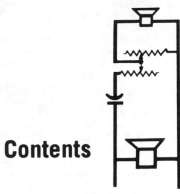

Contents

closures for Musical Instrument Speakers—Monitor
Speakers—Project 8: A Musical Instrument Speaker

Introduction

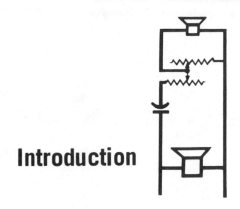

Welcome to the fun of speaker building. This versatile hobby, which has been growing in popularity for several years, combines many different skills. You get a bit of the sciences of electronics and acoustics, as little or as much as you want to dig into those areas. Then there is the practical art of furniture design, if that's your bent. And, of course, the fine art of music appreciation. Speaker building is an endeavor that makes as much use of art as of science.

This book contains many detailed project plans, from single speaker-in-a-box jobs to two and three-way systems. But, more important to those who like a challenge, you will find practical information on how to design your own system. Simplified design charts make it easy for the novice, but there are calculator and computer programs for those who want to further pursue the design process.

As a home speaker builder, you can design a system that looks and sounds the way you choose, instead of being restricted to what is available from the mass production line of a factory. A manufacturer must appeal to a mass audience; you have only yourself, and perhaps your family, to please.

In talking to many potential speaker builders I have found that the most common reason they have given for holding back is that they doubt their ability to complete a successful project. In almost every case a little encouragement convinced them that their fear was groundless. Many builders have begun with doubts—and ended with success. You can, too.

1

How a Speaker Works

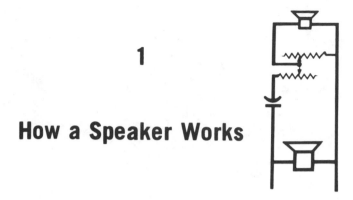

The speaker is the part of a sound system that has the fewest specifications, but the greatest effect on quality. Most listeners can't hear the difference with a change of amplifiers, but can immediately sense a switch in speakers. If you want to upgrade your stereo set, the speaker system is a good place to start.

PARTS OF A SPEAKER

The working parts of a dynamic speaker are the cone with its suspension, the voice coil, and the magnet (Fig. 1-1). When an electric current flows through a wire, it sets up a magnetic field around that wire, and for a coiled wire the field is increased. If the coil of wire is located in an external magnetic field, provided by a magnet, the field of the coil interacts with that of the magnet to apply force to the coil. If the current is an alternating current, the field around the coil builds up and collapses in response to the frequencies of the current. In a speaker this changing field interacts with the constant field of the magnet, causing the coil to move in response to the current. As the voice coil moves, it moves the cone, which makes pressure waves in the air near the cone. These pressure waves are heard as sound.

The cone suspension allows the cone to move without shifting sideways far enough for the voice coil to rub the center pole of the magnet. Modern speakers have an outer cone suspension, the *surround,* and a coil centering device called the *spider.*

If you connect the leads of a sensitive AC voltmeter to the terminals of a dynamic speaker and gently push the cone with your

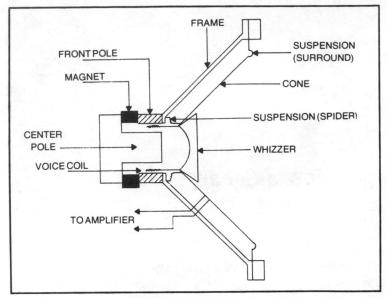

Fig. 1-1. Parts of a loudspeaker.

hand, the meter pointer will move, indicating that you have produced a small voltage across the terminals. When a speaker is connected to the input of a high gain amplifier, it can act as a microphone. Many intercom sets use this principle, with a small speaker in each station that works in a dual role of speaker during the receiving mode and microphone during the sending mode.

When a speaker is in use, each voice coil movement produces a voltage in the same way. The direction of the electromotive force (emf) produced is always that which will oppose the change of current direction in the current that drives the speaker. Because of its direction, this force is called the *back emf*. The stronger the magnetic field, the greater the back emf and the higher the electrical damping on the voice coil and cone movement.

FREQUENCY RESPONSE

In selecting a speaker for music reproduction, smoothness of response is more important than extended range alone. A speaker with a smooth response from 100 to 10,000 Hz can produce music more faithfully than one with a 50 to 15,000 Hz range but with significant peaks within that range.

The requirements for a wide frequency range from a single cone are contradictory. For good performance above 10,000 Hz the cone must be light, with a mass no greater than about 5 grams. But light cones bend at low frequencies, producing distortion. For good bass response the cone should be large to "grab" enough air to have good radiation resistance, but for good dispersion at high frequencies it must be small. A heavy bass cone will have a lower frequency of resonance and a smoother response than a light one, but the heavy cone will have poorer transient response as well as a limited high range.

One way that speaker designers solve these problems in full range speakers is to add a secondary cone, called a *whizzer*. A whizzer cone is less expensive than a separate tweeter because it is driven by the same magnet and voice coil as the main cone. It can improve the radiation pattern of the highs as well as extend the frequency range.

Even those large cones that have no whizzer can produce some high frequency sound. They do this by cone *reduction*. At low frequencies the entire cone vibrates as a single piston, but as the frequency of the signal is increased, the central part of the cone vibrates independently. Having lower mass, a small section of the cone can emit sound at relatively high frequencies.

TRANSIENT RESPONSE

A speaker's ability to handle short pulses without altering their duration is called its *transient response*. For good transient performance a speaker must start to move almost immediately after receiving the amplifier's signal to do so, then stop promptly when the signal ends. The trait of oscillating after the signal has ended is called *hangover*.

The first requirement for good transient response is a smooth frequency response. A peaky response curve indicates multiple cone resonances, and each resonance can be kicked off by any signal which contains the resonance frequency in either the fundamental tones or in its natural overtones. Each resonance adds its share of hangover, and the aural effect of hangover is muddy sound.

Even if a speaker has a smooth frequency response without higher resonances, it will have at least one resonance, the fundamental cone resonance. The prominence of this resonance varies with system Q; a high Q speaker has low magnetic damping and is prone to peaking at its frequency of resonance.

DISPERSION

All speakers produce sound that is more directional at the upper end of their frequency range. As a rule of thumb a speaker is omnidirectional only up to the frequency where the effective cone diameter is equal to the wavelength of the sound. Following this rule, a 12" speaker is fully omnidirectional to about 1300 Hz, an 8" speaker to 2000, or a 4" speaker to 4000. Speakers can be used to perform at higher frequencies than these because of cone reduction and the use of whizzer cones.

A speaker that has poor dispersion at high frequencies will sound harsh when you are in the beam but dull when you move aside. If such speakers are used in a stereo set, the stereo image may move unnaturally as you turn your head. A speaker that spreads the highs around the room will sound more expansive and the highs will have an airy quality like those of live music.

For the ultimate in good dispersion, small dome tweeters are hard to beat. The dome shape permits the necessary strength for a small vibrating surface, and the small size provides the superior dispersion.

CONE RESONANCE

If you suspend a mass on a spring (Fig. 1-2) and set the mass in motion, it will always vibrate at a certain frequency which is its natural *resonance*. To change the frequency of resonance you can alter the mass or the stiffness of the spring. If you add a blob of modeling clay to the mass, it will vibrate more slowly. Or a more compliant spring will have the same effect. To increase the frequency you would either reduce the mass or get a stiffer spring.

Each speaker has a fundamental frequency of resonance which is determined by the mass of the cone and the compliance of its suspension. Large cones, having greater mass, usually have a lower frequency of resonance than small cones. When the frequency of resonance is measured on a bare speaker, it is called the *free air resonance.*

If you sprinkle some talcum powder on a speaker cone and watch the powder as you vary the frequency of the drive signal from an audio generator, you will see that the cone's vibration increases as you approach the speaker's resonance. At resonance it vibrates wildly. At this frequency the speaker is extremely efficient at converting electrical energy into sound, but, by its greater voice coil movement, the speaker produces more back emf at this frequency than at any other. The stronger the magnet, the higher the opposition, or *impedance,* to the flow of current through the coil. This is why a strong magnet controls the cone movement at resonance, damping it.

COMPLIANCE

The traditional suspension, which was no more than a single wrinkle around the circumference of the cone, has been largely replaced by a roll edge in high fidelity speakers. High compliance woofers, with roll edge suspensions, can perform well in compact boxes.

Another change in speaker compliance is the way in which the compliance is reported. Instead of the distance the cone moves per unit of applied force, it is reported as that cubic volume of air which has the same compliance for the cone as the speaker's suspension. When reported as a cubic volume, the term is called the V_{AS}. Because any volume of air will offer more resistance to the movement of a large piston than that of a small one, large woofers almost invariably have a high V_{AS}. Before one can say whether a certain box size is large or small for a given speaker, the speaker's V_{AS} must be considered.

DAMPING

The degree of damping for any speaker depends on several factors, the size of the magnet, the suspension, the mass of the moving parts, and the internal resistance of the amplifier. If other factors are equal, an increase in magnet size produces more damping. But the rule "the bigger the better" doesn't go; too much magnet can overdamp a speaker, reducing the level of bass response. Magnet size isn't everything, but you can get an

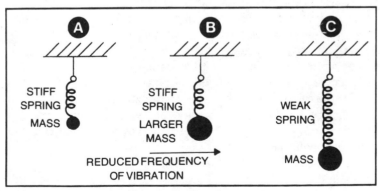

Fig. 1-2. How changing mass or compliance affects resonance.

indication of speaker quality by the size of the magnet simply because the magnet is the most expensive part of a speaker. Manufacturers aren't likely to squander the cost of a big magnet on a poor speaker. For optimum bass performance the magnet must be fitted to the size of speaker and the kind of enclosure to be used.

As a measure of the degree of damping on a speaker a numerical term, called speaker Q is used. Q stands for resonance magnification, the tendency of the speaker to peak in response at the frequency of resonance. Values for Q can range from about 0.2 up to 2 or 3 or more. The greater the damping on a speaker, the lower is its output at resonance and the lower is its Q. When a speaker is put into a closed box, its Q, as well as its frequency of resonance, will be raised. It is often accepted that a Q of 1 is a useful design goal for closed box speakers. This is a good compromise between a system that is under-damped and one that is over-damped.

IMPEDANCE

The opposition of the voice coil to current flow at any frequency is called its *impedance*. Most speakers have a nominal impedance rating of 8 ohms, which suggests that a speaker is like a resistor. The units of resistance and impedance (ohms) are the same, but there are important differences between the impedance of a speaker and pure resistance.

If you measure the resistance of your speaker's voice coil with an ohmmeter, you will find that it is about 75% of the rated impedance. An 8-ohm speaker, for example, will usually have a resistance of about 6 ohms. This tells you that impedance is something more than simple resistance. The ohmmeter measures resistance by putting a *direct* current through the voice coil, but the speaker must operate on *alternating* current. When an alternating current passes through the coil, the constantly swinging flow sets up its own magnetic field which grows and collapses with the frequency of the current. This moves lines of force through the coil, causing a reactance to the alternating current. The more rapidly the current reverses itself, the greater the reactance. This tendency of a coil to resist the flow of high frequency current is called *inductive reactance*. The more turns of wire in the coil, the greater its inductance.

A coil can also have *capacitive reactance*, which produces the opposite effect of inductive reactance. At frequencies where the capacitive reactance is equal to the inductive reactance, the two

14

reactances cancel, and the speaker's impedance is equal to the DC resistance of the voice coil. At all other frequencies the total reactance to current flow will be greater.

Impedance varies with frequency (Fig. 1-3). Note the hump at the frequency of resonance where the voice coil's back emf is greatest. The rise at high frequencies is caused by voice coil inductance. While the uneven impedance curve may look bad, remember that this is not a response curve. The high peak at resonance is a sign of a strong magnetic field.

When connecting speakers in your stereo system, the important thing to remember about impedance is to avoid a connection that puts a low impedance load on your receiver or amplifier. For many components the danger line is approached if the impedance drops below 4 ohms. Just make sure you don't wire more than two 8-ohm speakers in parallel. And never wire 4-ohm speakers in parallel with any other speakers. Many receivers and amplifiers have speaker switches that put the main speaker system and a secondary set of speakers in parallel when both sets are on, so you should be careful about using 4-ohm speakers in such combinations.

EFFICIENCY

The measure of any speaker's ability to convert electrical energy into sound energy is its *efficiency*. Mathematically, efficiency is sound power output (in acoustical watts) divided by electrical power input (in electrical watts). There are two ways to

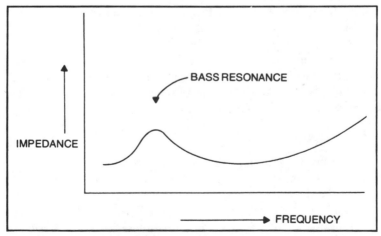

Fig. 1-3. How a speaker's impedance varies with frequency.

increase efficiency: reduce the cone's resistance to motion, or increase the force of the voice coil for a given flow of current.

The easiest way to reduce the cone's resistance to motion is to make it lighter. If the cone mass is halved, the speaker will be 4 times more efficient in the mid-frequency band. But reducing cone mass increases the frequency of resonance and often produces a rougher response curve. Another way to reduce motion resistance is to increase the compliance of the suspension.

To increase the force on the voice coil, the manufacturer can use a stronger magnet or put more turns of wire in the coil. Careful design must be used with these methods to prevent overkill. Too much magnet will over-damp the cone, and if too much wire is added to the coil, the increase in mass will begin to cut efficiency more than the field of the added turns can increase it.

Efficiency is one speaker characteristic that is rarely quoted. For many purposes it is not especially important because of the available power from modern receivers and amplifiers. It may be of importance to listeners who like extremely high sound levels.

Compact, closed box speakers are the most inefficient kind, with efficiency ratings of from 0.25 to 0.5%. Large floor models often have much higher efficiency, sometimes as high as 20 times those of the most compact speakers.

Another way of rating efficiency is to show the sound pressure level (SPL) in decibels (dB) when the speaker is fed 1 watt of electrical energy. This measurement is made with the microphone at 1 meter from the speaker. If you check such ratings, you will see that small speakers and tweeters almost invariably show higher dB ratings than large woofers. This is why tweeters and mid-range speakers often need either specially designed crossover networks or variable controls for proper balance. Such dB ratings should be used as a rough guide and have almost no value in rating speaker quality. A heavy cone woofer, suitable for use in a compact enclosure, will usually seem inefficient when compared to full-range speakers of the same size.

POWER RATING

The power rating that most manufacturers assign to a speaker is the amount of power the speaker can absorb without damage. Some of the electrical energy that goes into a speaker's voice coil is converted to heat by the coil's resistance. The larger the coil, the better it can dissipate heat, so you can estimate the power handling ability of a speaker by checking the diameter of its voice coil.

Speakers with the smallest voice coils, from ½" to 9/16", are usually rated at no more than 5 watts. The power ratings rise with increased coil diameter, so that a speaker with a 1" voice coil can handle from 15 to 30 watts. Speakers with coils larger than 1" can usually handle much more power. A 2" voice coil speaker, for example, can be rated at 100 watts or higher.

These ratings are rms values, which means that you can use an amplifier with a much higher power rating than the speaker. Music is full of transient sounds rather than sustained tones. When music power ratings are quoted, the figure is always considerably higher than the rms figure.

In some cases an amplifier or receiver with a high power rating is safer for your speakers. Low powered amplifiers can produce distortion that adds upper harmonics, placing an undue load on small tweeters. Ordinary music contains little power in the high frequency range, but a poor 10-watt amplifier can blow a tweeter which would normally be perfectly safe in a 20-watt or even more powerful system. By going into harmonic distortion at transient peaks, the low powered amplifier tremendously increases the power to the tweeter from the normal milliwatt range into several watts.

SPEAKER POLARITY

If you hook up a single speaker to a mono amplifier or receiver, it makes no difference which speaker lead goes to each terminal. But when two or more speakers cover the same frequency range in the same room, they must push and pull together or the sound from one will cancel that of the other (Fig. 1-4).

Stereo speakers have coded speaker terminals. The positive terminal is usually marked with a red dot, sometimes with a plus mark. Make sure that the lead from the positive terminal of each stereo speaker goes to the receiver terminal of the same polarity in each channel.

If you think your speakers may be out of phase, try reversing the leads to one speaker. Note that you must reverse the wires to just one speaker; if you reverse the wires to both stereo speakers, their polarity with relation to each other will be the same as in the first wiring. The easiest way to note proper polarity is to place the two speakers face to face and feed a mono signal with prominent bass tones to them. Switch the connections on one speaker only. Choose the connection that gives the greatest bass response.

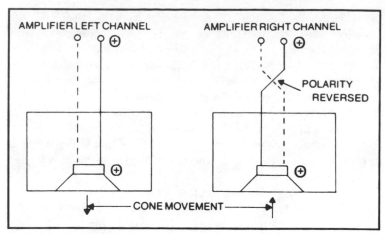

Fig. 1-4. Why stereo speakers must be in phase.

It is also important to observe polarity in wiring together woofers and tweeters in 2-way systems or the three drivers in 3-way systems. Failure to do this can produce holes in the response curve. Unless you are instructed to do otherwise, the various drivers should be wired in phase. Some crossover networks, such as second-order 12 dB per octave networks, produce a phase difference in adjacent drivers of 180° and require reversed phase wiring to prevent a hole in the response curve. If reversed polarity is necessary, it will be indicated in either the instructions or the schematic diagram.

KINDS OF DRIVERS

Any direct radiator speaker is called a *driver*. Because of the conflicting demands placed on a single cone driver for full-range duty, nearly all systems with a driver larger than 8" in diameter are multiple speaker systems. These systems have a large driver for the low frequencies, often called a *woofer,* and at least one other driver for the highs, a *tweeter*. A 2-way system consists of a woofer and tweeter; a 3-way system of a woofer, mid-range driver, and tweeter.

Most woofers have an advertised diameter of at least 8 inches. If a driver is designed to be a woofer, it will have limited high frequency response. This limitation is desirable because it makes the crossover network's job easier. A woofer will also have a low frequency of resonance, low enough that its bass range will be adequate when the speaker is put into a suitable box.

18

Mid-range drivers are often considered unimportant on the theory that any speaker can reproduce mid-range. This attitude is a mistake because the ear is most sensitive to response variations in the mid-range. For best mid-range performance, choose a driver that is designed for the purpose. And if your woofer is larger than 8" or 10" in diameter, you will probably need a mid-range driver for good mid-range performance.

The best tweeters are usually small in diameter, giving better dispersion than large tweeters. This is no problem in 3-way systems; the tweeter can be a 1" dome. But small tweeters must be used with adequate crossover networks. If the crossover frequency is placed too low, or if the tweeter is incorrectly wired, it won't last long. Remember that the power ratings assigned to tweeters are based on the assumption that an adequate crossover network will be used.

Piezoelectric tweeters can solve many problems of high frequency reproduction. These tweeters have such a high impedance at low frequencies that they are effectively out of the circuit there. This means that they can be used without a crossover network, wired directly to the amplifier output line. Such tweeters are unusually rugged, able to take driving voltages of up to 35 volts rms without failing. For continuous high power level operation a current-limiting resistor is suggested because these tweeters have low impedance at frequencies above 50,000 Hz.

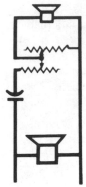

2

Kinds of Speaker Enclosures

A speaker in free air is like a fish out of water. Tests on unbaffled speakers vs. properly enclosed speakers show that after installation in a suitable box a speaker can deliver up to 100 times greater sound intensity at low frequencies than it can in free air. The unbaffled speaker's lack of ability in bass performance explains why its sound is so thin and unbalanced.

FUNCTIONS OF ENCLOSURES

To see why a bare speaker sounds bad, consider Fig. 2-1. The plus signs represent an increase in pressure as the cone moves against the air; the minus signs, a decrease. When air from the high pressure side of the cone mixes with air from the low pressure side, sound cancellation occurs. At high frequencies the sound is directional, so little mixing occurs, but at frequencies where the wavelength is long compared to the diameter of the speaker, the waves can curve back around the cone so that the out-of-phase waves mix. One of the basic requirements of a speaker enclosure is that it block this unwanted mixing of out-of-phase waves.

A baffle or enclosure does more for a speaker than merely prevent cancellation. When an unbaffled speaker tries to pump air, it meets little air resistance, so, like a piston out of a car engine, it can do little work. Any speaker, large or small, can work better if it is installed in a baffle or box so that the air cannot get away so easily when the cone moves against it.

Another function that various kinds of enclosures serve, in different degrees, is to damp the speaker. Hangover kills definition, the speaker's ability to define individual tones and instruments. A mismatched enclosure can aggravate hangover, but a properly designed box will reduce it to a minimum. When a speaker is installed in a box, the air pressure against the cone adds an acoustical resistive load to it. Like the damping pads on a piano, this box air pressure damps excessive cone movement. An unbaffled speaker has electrical damping but little mechanical damping.

TYPES OF ENCLOSURES

There are four kinds of speaker enclosures in general use: closed boxes, ported boxes, labyrinths and their variations, and horns. A fifth type, the flat baffle, is found occasionally. The first two types, closed boxes and ported boxes, are used for the great majority of stereo speaker systems.

Closed Box Enclosures

When a speaker is installed in a closed box, the air in the box acts as a spring against the cone, particularly if the box is small. The smaller the box, the stiffer the air spring. For high compliance speakers, labeled as "air suspension" or "acoustic suspension" types, the box air usually provides more restoring force to the moving cone than the suspension. One of the arguments for such speakers, when they first appeared, was that the air behind the cone provided a more linear restoring force than the stiff suspensions used in earlier speakers.

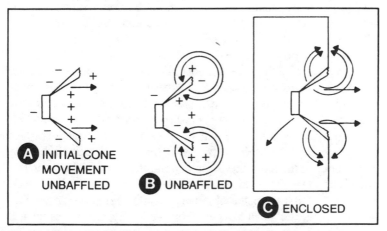

Fig. 2-1. Why an unbaffled cone is inefficient at low frequencies.

To take full advantage of the compact closed box principle, air suspension speakers must be specially designed. They need longer voice coils so the cone can move farther without carrying the coil too far from the magnet. The cones are also made heavier, which reduces the frequency of resonance and makes for a smoother frequency response curve.

After analyzing earlier large closed box speakers, the engineers who developed the compact acoustic suspension speakers made a clever trade-off. They swapped a large high compliance box for a small box with low air compliance, then to maintain low frequency bass range they replaced the low compliance stiff cone with a loose cone suspension (Fig. 2-2). Note the combinations: loose (large) box, stiff speaker; stiff (small) box, loose speaker. Almost everyone likes to save space, so the popularity of the compact closed box came as no surprise.

Like all other kinds of speakers, small closed box systems have their disadvantages. The most frequently mentioned one is lack of efficiency. Another weakness, the longer voice coils and greater cone mass in high compliance woofers put a strict upper limit on the woofer's frequency response. This means that with these woofers you must either use a tweeter that will extend well down into the mid-range or go to a 3-way system.

Ported Box Enclosures

Like the closed box, the ported box, or *bass reflex*, is easy to build; unlike the closed box, its action is complex. If you cut a small hole in a closed speaker box, the air in the box retains its ability to act like a spring, while the air in the hole acts like another piston. This air piston vibrates in phase with the cone at some frequencies, out of phase at others (Fig. 2-3).

The box with the hole in it acts as a resonator, properly called a *Helmholtz resonator*, after the nineteenth-century German physicist who first described the behavior of tuned acoustical resonators. An empty pop bottle acts as a Helmholtz resonator when you blow across the open end of the bottle. The frequency of resonance for any Helmholtz resonator is determined by the compliance of the air in the container and the mass of the air in the port. Like the speaker itself, a ported box is a tuned circuit in which a mass resonates against a compliance. At the Helmholtz frequency of resonance, the air in the port vibrates easily, compressing and decompressing the air in the box.

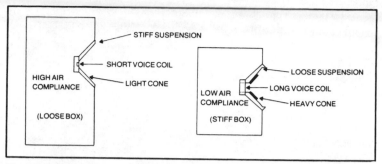

Fig. 2-2. How speaker type affects closed box size.

When a speaker is installed in a ported box, and is closely coupled to the tuned circuit of the box, the original speaker resonance is replaced by two new resonances, one at a higher frequency than the original, the other at a lower frequency (Fig. 2-4). At the upper resonance the air in the port moves in phase with the cone, but, because this frequency is well above that of the box resonance, the damping action is reduced. This is the frequency at which some reflex systems produce too much output, particularly small enclosures whose upper resonance peaks occur within the frequency range of the male voice.

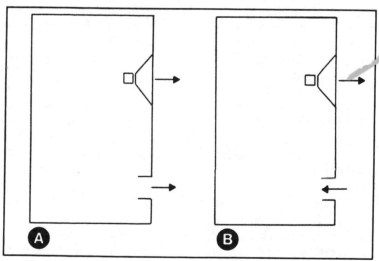

Fig. 2-3. At the frequency to which the box is tuned the port air moves in phase with the speaker cone, damping it (A). At ultra low frequencies the port air no longer moves with the cone and the speaker may be unloaded. This disadvantage of ported speakers can be relieved by use of the infrasonic filter on the receiver or amplifier.

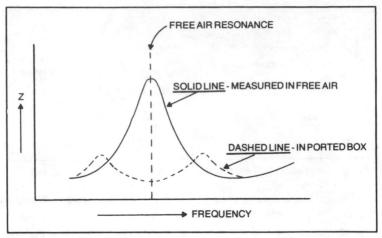

Fig. 2-4. How the impedance curve of a speaker changes when it is placed in a classic bass reflex enclosure.

If the box is properly tuned, the vibrating air piston works in phase with the cone over a selected band of frequencies to reinforce bass output. This action is most effective at the Helmholtz frequency. Here the port air acts to damp the cone so that it hardly moves, although the port air is moving at maximum velocity. In the "classic" bass reflex system illustrated by Fig. 2-4 the box is tuned to the speaker's free air resonance. Note the dip in the impedance curve at that frequency, indicating restricted cone movement. This shows the damping action of the port. Because distortion varies directly with cone movement, this cone damping action reduces distortion.

Below the box resonance frequency, there is a rapid phase shift in the output of the port so that at the lower resonance the port radiation is 180 degrees out of phase with that of the cone (Fig. 2-3). The out-of-phase radiation, plus the normal roll-off in speaker response at ultra low frequencies, produces a low bass cut-off rate much sharper than that of closed box speakers.

To tune a ported box you can change the area or the length of the port. Compact ported boxes are usually tuned by a ducted port because the use of a duct increases the mass of the vibrating air, tuning the box to a lower frequency. If the box has a simple port with no duct behind it, you can increase the mass of the vibrating air by making the port smaller. While this may seem surprising, a smaller port increases the air velocity and permits the port air to carry more air with it.

When the audio world first became aware that a tube behind the port made it possible to tune small boxes to lower frequencies, this discovery was greeted with enthusiasm in the belief that it permitted small enclosures to give a bass range and level equal to that of large ones. This idea confused tuning with total performance. It neglected the important relationship of speaker compliance to box air volume compliance. The only small reflex enclosures that can outperform larger ones are those with a compliance and tuning that is optimum for a specific driver.

One of the peculiarities of a reflex system is that port radiation does not vary with the size of the port. A small port will radiate just as much sound as a large port, but at higher velocity. If the port is too small, it can produce unmusical noises by its high air velocity. This weakness of small reflex enclosures can be solved by substituting an extra diaphragm, usually called a *passive radiator*, for the air in the port. A passive radiator is made like a woofer but has no magnet or voice coil. It can be tuned by changing the mass of the vibrating diaphragm.

The most important advantage of the ported box is its damping control on the speaker. It also offers more efficiency by permitting a more efficient woofer. A woofer designed for a ported box does not have to have as heavy a cone or as long a voice coil as a sealed box woofer, so it can have more extended mid-range response. This makes a 2-way woofer-tweeter system a practical possibility.

The biggest disadvantage of the reflex is the complexity of speaker to box relationships, requiring careful design for good performance. A less obvious disadvantage is its tendency to unload the speaker below resonance. Here, turntable rumble or other subsonic pulses can overload the speaker, causing distortion or even damage. The cure is to make sure that any receiver or amplifier to be used with a ported system has a low-cut filter to remove low frequency garbage from the signal. Such filters don't remove useful bass; instead they make the bass more solid by removing unheard pulses that can stress the amplifier as well as the speaker.

Labyrinths and Transmission Lines

A labyrinth is a tuned pipe with the driver at one end and a port at the other end (Fig. 2-5). When the wave from the driver reaches the end of the pipe, it expands into the room, causing a sudden pressure drop. This drop in pressure is reflected back through the pipe to the speaker. At the frequency where the length of the pipe is equal to a quarter wavelength ($\lambda/4$) of the sound, the air at the

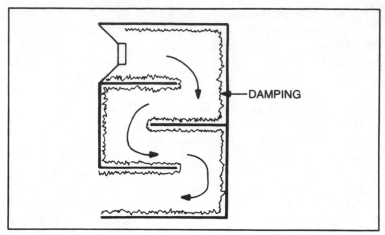

Fig. 2-5. Labyrinth.

mouth of the pipe is at minimum velocity, but maximum pressure. At that frequency, the change of pressure is also at a maximum, and the reflection to the speaker of the rarefaction produces maximum damping. A labyrinth that has a length equal to one quarter wavelength at the speaker's resonance frequency will damp the speaker at resonance, an action similar to the bass reflex. Labyrinths usually have a series of resonant peaks, occurring at harmonics of the fundamental pipe resonance, but these can be subdued by lining the walls with damping material.

A transmission line is a stuffed labyrinth (Fig. 2-6). The theory behind the transmission line was developed by A.R. Bailey, a professor in England. Bailey suggested that reflex enclosures, by their sharp cut-off rates, were likely to produce ringing. He reasoned that an acoustic line behind the speaker that was infinitely long would absorb the backwave without causing troublesome reflections, but to keep the line within practical limits, he substituted stuffing for length. Extremely low frequency waves are not absorbed by the stuffing but emerge from the mouth of the line to augment the speaker's low bass response. While labyrinths usually have a cross-sectional area that is equal to the speaker cone area, transmission lines are often tapered to spread the resonances. The large end is usually greater in area than the cone, the port end smaller.

Transmission lines have obvious disadvantages. They are large, require a complicated structure, and can be unpredictable. Fine tuning a transmission line is largely a matter of trial and error.

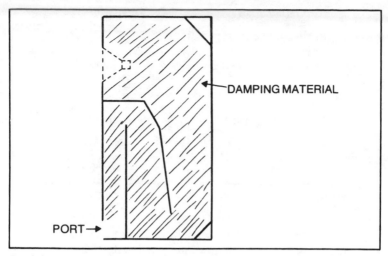

Fig. 2-6. A transmission line enclosure. It differs from the labyrinth by its taper and stuffing.

Horns

A horn acts as an acoustical transformer. It matches the high acoustical impedance at the driver to the low impedance of the room air by its smooth rate of increased cross-sectional area from the driver cone to the horn mouth (Fig. 2-7). An acoustical megaphone, the kind used by cheerleaders, has some of the virtues of the horn. However, being straight sided, it does not permit the sound waves to expand at a constant rate. As the area of a megaphone increases, the distance between the points where the area doubles also increases. A true horn has a flare which forces the sound waves to expand at a constant rate. Because of its impedance-matching characteristics, a horn offers much higher efficiency than other types of speaker enclosures. The horn permits a lower distortion at high output than any other enclosure because of its high damping on the driver.

The disadvantages are obvious. The size required for a bass horn is tremendous; some are 30 feet long. Folded horns are

Fig. 2-7. Straight exponential horn.

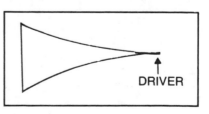

Table 2-1. Summary: Closed Box vs. Reflex.

ADVANTAGES OF CLOSED BOX	ADVANTAGES OF REFLEX
1. SIMPLE TO DESIGN AND BUILD	1. HIGHER OUTPUT AND LOWER DISTORTION IN OCTAVE ABOVE f_b
2. MORE GRADUAL CUT-OFF RATE OF 12 dB/OCTAVE	2. GREATER BASS RANGE OR HIGHER EFFICIENCY FROM EQUAL BOX VOLUME
3. GOOD FOR SUB-MINIATURE ENCLOSURES	3. CAN USE SINGLE CONE SPEAKER WITH REDUCED PHASE PROBLEMS
4. CAN USE WITH HIGHER POWERED AMPLIFIERS BECAUSE IT DOESN'T UNLOAD SPEAKER AT LOWEST FREQUENCIES*	4. LOWER COST BECAUSE OF #3
5. SPEAKER VARIABILITY HAS LESS EFFECT ON PERFORMANCE	5. GREATER CHALLENGE— GREATER REWARD IF DONE RIGHT

*THIS ADVANTAGE FOR THE CLOSED BOX CAN BE NEUTRALIZED IF AN EQUALIZER OR INFRASONIC FILTER IS USED WITH REFLEX SYSTEMS.

sometimes used in home stereo systems, but because of their size and complexity, horns are found chiefly in theater sound systems or other high level applications.

WHICH ENCLOSURE IS BEST?

Even if one of the enclosure types described here could be proven to give the best performance in every system, it would not automatically become the universal choice. The bass horn is a good example of a high performance enclosure that is made rare by its cost and complexity. Table 2-1 summarizes the chief advantages of the two most commonly used types of enclosures.

Table 2-2. Parts List for Project 1.

¾" Plywood or Particle Board:	
2 8" × 17¼"	Sides
2 8" × 11¼"	Top & bottom
2 9¾" × 15¾"	Speaker board & back
¾" Pine:	
4 ¾" × ¾" × 15¾"	Side cleats
4 ¾" × ¾" × approx. 8¼"	Top & bottom cleats
⅜" Plywood:	
1 9⅝" × 15⅝"	Grille board
¼" Hardboard:	
1 7" × 8"	Crossover board
Speakers and Components:	
1 6½" woofer	Radio Shack Cat. No. 40-1009
1 Cone tweeter	Radio Shack Cat. No. 40-1270
1 8-ohm L-pad	Radio Shack Cat. No. 40-980
1 4 μF capacitor	(supplied with tweeter)
Miscellaneous:	
Fiberglass insulation	Radio Shack Cat. No. 42-1082
Grille cloth	

PROJECT 1: A COMPACT TWO-WAY SPEAKER

This project is built around a 6-½" woofer and a small cone tweeter (Table 2-2). It is easy to build and wire, requiring only a capacitor, supplied with the tweeter, and an L-pad. Although its bass response doesn't go as deep as that of the larger speakers in this book, it is smooth and uncolored.

This speaker cabinet is so small (Fig. 2-8) you can probably get left over scraps of plywood from a cabinet shop large enough to do the job (Fig. 2-9). If you have no easy way to make the 45° beveled cuts at the ends of the enclosure walls, you can use butt joints. Just make sure the internal dimensions are approximately the same as those shown in the plans.

Construction

Cut out the parts and set the enclosure shell, the four walls, together to test their fit. It is much easier to install the cleats that hold the speaker board and back panel before assembling the shell, especially in a small enclosure such as this one. Cut and install the side cleats, at least. One way to make them fit properly is to rope or strap the parts together and carefully measure the exact length of cleats needed. If the cleats are too long, they will interfere with further assembly; if too short, there will be gaps. Minor gaps, of about ⅛", can be filled with latex caulking compound or silicone

Fig. 2-8. Finished speaker is small enough for shelf use.

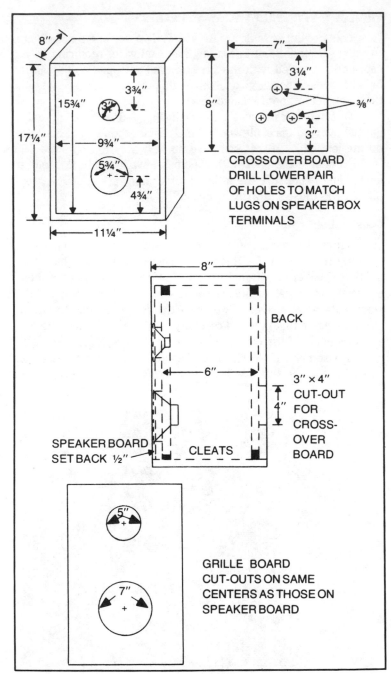

8"

17¼"

15¾"

3¾"

3"

9¾"

5¾"

4¾"

11¼"

7"

3¼"

8"

3"

⅜"

CROSSOVER BOARD
DRILL LOWER PAIR
OF HOLES TO MATCH
LUGS ON SPEAKER BOX
TERMINALS

8"

BACK

6"

3" × 4"
CUT-OUT
FOR
CROSS-
OVER
BOARD

4"

SPEAKER BOARD
SET BACK ½"

CLEATS

5"

7"

GRILLE BOARD
CUT-OUTS ON SAME
CENTERS AS THOSE ON
SPEAKER BOARD

Fig. 2-9. Construction plans for Project 1 enclosure.

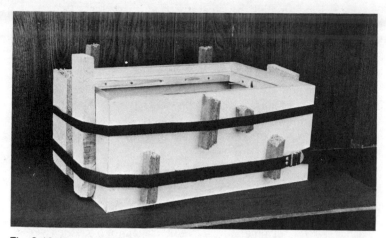

Fig. 2-10. Insert small blocks under straps in middle of each side, as in upper strap, then move the blocks toward corners of enclosure, as in lower strap, to apply pressure to corner joints while gluing.

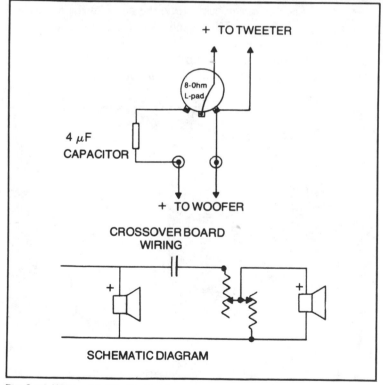

Fig. 2-11. Wiring circuit for Project 1.

rubber sealant. Install the cleats with glue and nails that are no more than 1¼" long. Note that you must leave a space of 1¼" at the front and ¾" at the rear between the cleats and the edges of the panels to accommodate the speaker board and back panel.

If you choose butt joints, use glue and small finishing nails. For the bevel jointed box, as shown, spread a coating of glue on the matching beveled surfaces and rope or strap the box together. To apply pressure to the joints, place small blocks of wood under the ropes or straps near the center of each side, then slide them toward the corners of the box (Fig. 2-10).

Drill the crossover board, according to the plans, and install a set of terminals on the outside with glue and screws. Radio Shack push-button terminals, Cat. No. 274-315, or spring-lever terminals, Cat. No. 274-621, are good choices for this project. Install an L-pad above the terminals. Remove the capacitor from the tweeter and install it between the positive board terminal and pin #3 of the L-pad. Wire the board according to Fig. 2-11. Install the board on the interior surface of the back panel, placing it over a 3" × 4" cut-out in the center of the back. Glue the board to the back with latex caulking compound or silicone rubber sealant; screws are optional unless you plan to work with it while the glue is setting. This completes the basic enclosure parts: the shell, the speakerboard, and the back panel with its crossover board (Fig. 2-12).

Fig. 2-12. Basic enclosure parts.

Fig. 2-13. Stretch grille cloth over face of grille board, then staple it on back of board.

Cover the front edges of the plywood enclosure walls with veneer and stain the box. Do all the finishing work, such as sanding and varnishing before installing the speakers.

Install the speaker board with glue and nails or screws. Caulk the interior joints; then glue and nail down the back. You can reach through the speaker holes to caulk around the back panel. Place the box on its back and loosely fill it with damping material, feeding the speaker leads out through the proper holes. Wire the speakers, soldering the connections. Then run a bead of silicone rubber sealant around each speaker hole, and press the speaker down on the sealant. Leave the box on its back overnight for the glue to set. Test the speaker by connecting it to a receiver and listening at low volume. Check to see that the treble comes from the tweeter and the bass from the woofer.

Make a grille board and paint the front flat black. Stretch some grille cloth over the front of the board, and staple it on the back (Fig. 2-13). Press the grille board into place. If it fits loosely, use hook and loop retainers.

Place the speaker in its permanent location and adjust the tweeter control. Turn the tweeter volume down, then bring it up until output from the two drivers blends so that you hear them as a single speaker.

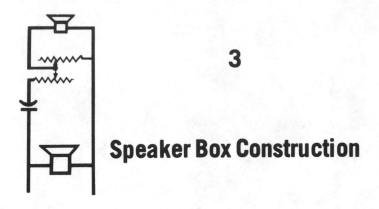

3

Speaker Box Construction

The principles of good speaker enclosure design and construction hold for any speaker project, regardless of box type. By heeding those principles you can build with confidence and prevent long debugging procedures. If you are inclined toward novelty in your choice of furniture, let that trait be expressed in your speaker enclosures by such superficial aspects of design as external appearance. For those design features that affect the sound, stick to basics.

DESIRABLE ENCLOSURE CHARACTERISTICS

The single most desirable characteristic of a good enclosure is that it does not seriously alter the sound of the speakers installed in it. A speaker system is different from a musical instrument in one important aspect: the speaker should produce no sound of its own. An instrument *produces* sound; a speaker *reproduces* it. If the speaker adds anything to the signal it receives, it is producing distortion. If the enclosure walls vibrate audibly, they will color the sound and weaken the true bass response by absorbing low frequency energy. A bad enclosure shape can spoil reproduction by lumping the inevitable air resonances around a narrow band of frequencies, producing a boomy system.

ENCLOSURE SHAPE

Enclosures with non-parallel walls, such as triangular boxes, have fewer problems with internal reflections. Most speaker builders as well as other audio fans choose ordinary box-like enclosures because they are easy to build and "look like speakers." To make the most of these, avoid extreme shapes. As a rule of thumb make sure that no internal dimension measures more than 3 times that of any other. But don't make them too close to each other or you will have a cube, one of the worst shapes for any but miniature boxes.

A preferred ratio of dimensions, frequently cited by acoustical engineers, is 0.62 : 1 : 1.62. This ratio, based here on its acoustical properties, is the same "golden ratio" of artists since the days of the Egyptian pyramids. Unless you have a special reason to use another ratio, this one will insure that the resonances will be spread instead of lumped. Figure 3-1 gives the right internal height, width, and depth for speaker boxes with net internal volumes ranging from 400 cubic inches (0.23 cubic foot) to 10,000 cubic inches (5.8 cubic feet). If you multiply the figures you get from the graph, you will find that the cubic volume obtained will be slightly larger than the one shown on the left margin. The extra volume of about 10% was calculated into the graph to allow space for speakers, cleats, and other internal parts.

Here is an example. Suppose you want to build a box with a net internal volume of 4000 cubic inches (2.3 cubic feet). Follow the horizontal line from 4000 in the left margin and find where it crosses each dimension curve, then read that dimension by referring to the values given at the bottom of the graph. Here we find the 4000 line crossing the D (depth) line at 10″, the W (width) line at about 16½″, and the L (length or height) line at 26¾″. Multiplying 10 × 16.5 × 26.75 we get 4413.75 cubic inches. So the overvolume here amounts to about 414 cubic inches, probably enough to allow for the internal space occupied by the speakers and internal bracing unless you are building a reflex cabinet that includes a large tube for a port. Where large ducts are used, the volume of the duct should be estimated and added to the required cubic volume. For closed box speakers the dimensions given by Fig. 3-1 will be satisfactory.

In some cases you must choose another ratio for various reasons. One example, if you want to install a 3-way system in a compact enclosure and want to mount the 3 drivers in a vertical line, which is a desirable pattern, you may have to make the cabinet's length to width ratio greater than 1.62 in order to make room for the speakers. In such situations try to stagger the dimension so that the width and depth are not equal.

Note that the dimensions given in Fig. 3-1 are internal measurements, so you must add enough length and width to the

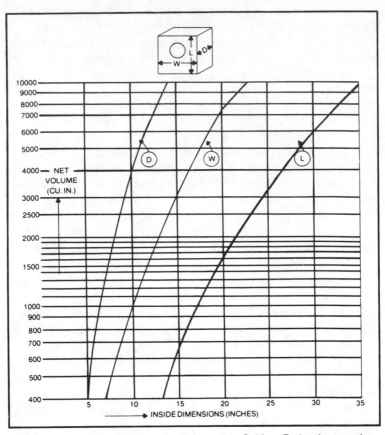

Fig. 3-1. Internal dimensions, based on the Golden Ratio, for speaker enclosures with a net cubic volume of 400 to 10000 cubic inches. These dimensions allow about 10% overvolume.

wall panels to account for wall thickness. For example, if you want to build an enclosure with bevel joints, such as you might use with a good grade of plywood, you would add 1½" to the internal length and width to get the proper length for side, top and bottom panels. If you add a particle board liner, as done with some of the projects in this book, you would add enough extra length to account for a double thickness of the liner. For an enclosure with butt joints you would add 1½", but only to those panels that overlap the other panels.

SPEAKER PLACEMENT

The first rule on driver placement is: put the high frequency speakers above the low frequency speakers (Fig. 3-2). If followed, this practice puts the tweeters nearer ear level, where their directional sound won't be lost, and woofers nearer the floor for good bass reinforcement.

The second rule is: put the drivers in a vertical line with little or no horizontal displacement. When speakers are mounted side by side, there will be a difference in path length from each speaker to the listener's ear, producing phase distortion. Sometimes lack of space makes side by side mounting inevitable.

The third rule is: if the vertical line of drivers is placed off center, make the second speaker board of your stereo pair a mirror image of the first. This allows you to place the speakers so that in both enclosures the drivers will be off centered inward, or outward, rather than having one enclosure with the drivers inward and the other with the drivers outward from the space between your speakers.

MATERIALS

The most commonly used speaker box material is ¾" plywood or particle board. Panel thickness should be chosen to match enclosure size. If the box is small enough, under ½ cubic foot in volume, it can be made from ½" or even ⅜" material. Small panels are stiffer than large panels of the same thickness, so you

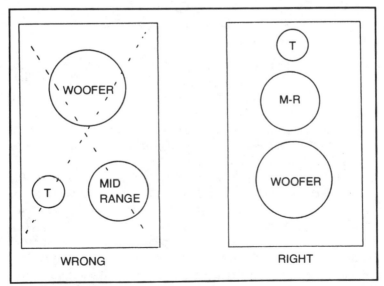

Fig. 3-2. Right and wrong speaker placement. Some designers prefer to mount the line of speakers off center if space permits.

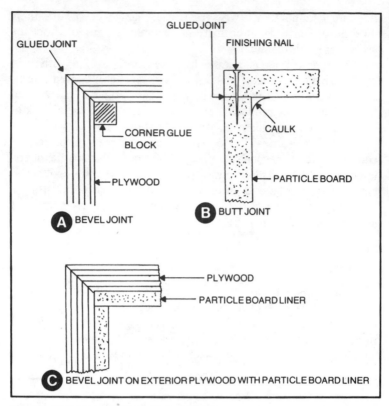

Fig. 3-3. Methods of joining enclosure panels.

should adjust thickness to panel size. A good rule of thumb is to use ¾″ material in enclosures made for woofers with a diameter of 8″ or greater. Brace large panels, of 2 to 3 square feet or more in area, to stiffen them.

If you use plywood, choose a good grade. Hardwood plywood, such as birch or walnut, is usually well made, but cheap grades of fir plywood will have many voids and even poorly glued layers. Voids weaken a panel, and loose layers can cause rattles. If you use fir, ask for "AB" grade of interior plywood. If you must use a cheap fir plywood, enlarge the enclosure by 1″ in each dimension and line the interior walls with ½″ particle board.

For a fine wood furniture appearance you can use ¾″ hardwood plywood with beveled corner joints (Fig. 3-3A or 3-3C), or you can build the box from particle boards with butt joints (Fig. 3-3B), then apply a wood veneer. Or you can cover the box with any kind of material you choose, from cloth to wallpaper.

Many audio fans would like to build speaker enclosures but lack power tools and feel that their homemade boxes would look amateurish. If this is your situation, you can make use of a design that permits a wrap-around grille cloth. The grille cloth will hide all matching joints, making your cabinet work look good even if it is rather crude. The basic procedure is shown in Fig. 3-4. You can build the particle board box, then add slab sides that extend beyond the box, and finally install the grille cloth. If you prefer a hard surface on the top of the enclosure rather than grille cloth, you can install boards top and bottom and wrap the grille cloth around from side to side as shown in Fig. 3-5.

Pine can be used for the exterior walls of the enclosures, but it isn't as strong as plywood and much less dense than particle board. It is particularly useful for slab sides in the easy-to-build enclosure of Fig. 3-4. Where it is used in a more conventional way, such as in Project 8 in Chapter 10, the box should be lined with ½" particle board. Pine is attractive but must be handled carefully to prevent it from warping. One precaution to observe, don't cut out the parts until you are ready to assemble them.

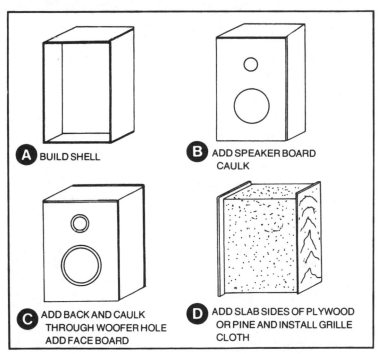

Fig. 3-4. How to construct speaker enclosures without use of power tools.

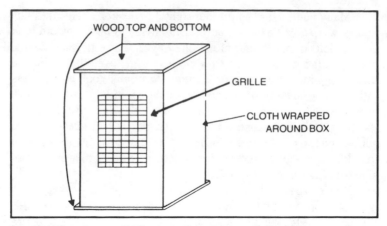

Fig. 3-5. Alternate way of hiding box with wrap around grille cloth.

Particle board is especially desirable because of its high density, lack of voids or loose layers, and its low cost. Its only disadvantage is that it dulls saws more quickly than other materials mentioned here. If possible use a separate saw, or saw blade, for cutting particle board. When buying it, choose an industrial grade. You can identify the most satisfactory particle board by examining the particle size at the edges of the board. The best board for speaker boxes has small particles and smooth edges. Avoid the kind that is crumbly or flaky.

CONSTRUCTION RULES

There are many ways to build a satisfactory speaker enclosure. You may want to change some of the construction methods shown in the projects in this book; just make sure you don't make significant changes in cubic volume. In building any speaker box, observe the accepted rules of construction.

●**Rule 1: Make A Rattle-Free Box.** The best way to avoid rattles is to do a good job of gluing the joints. Experienced carpenters use just enough glue to hold, but not enough to squeeze out onto raw wood and spoil the appearance. That technique works well with most furniture but, if overdone, can cause problems in speaker boxes. Unless the parts are cut to close tolerances, the skimpy use of glue can leave air leaks as well as rattles. To avoid problems with excess glue, cover the exterior wood near the joint with masking tape before gluing. And spread the glue thicker near the interior side of the joint. After assembling the parts, remove excess glue with a damp cloth.

If you make one panel removable, use carefully installed foam weather stripping material to seal the box and prevent rattles. Such panels will usually be held in place with wood screws, size #8 × 1¼" is about right, screwed into solid wood cleats. Position the weather stripping so the screws won't catch and damage it. When installing removable panels, place the screws no farther apart than about 4 or 5 inches.

●**Rule 2: Make The Box Airtight.** Even the smallest gap between joints can cause air noises that mar performance. If the air leak is significant, it can unload the woofer, increasing distortion and limiting its power handling ability. To insure against leaks, caulk every joint with latex caulking compound. If you are building a box with no removable panels, you can start by assembling the sides, top, bottom and the speaker board. Caulk every inner joint (Fig. 3-6), then install the back. You can reach through the speaker holes to caulk the joints around the back.

Apply the caulking material with a gun; then, if necessary, smear it into the joint with a rounded tool. Popsicle sticks make good spreading tools.

●**Rule 3: Don't Forget Damping Material.** Speakers that don't have enough damping material sound loud, even at low volume. Damping material is necessary to absorb sound from the rear of the cone—sound that would otherwise be bounced around the box and reflected out through the cone. Damping material suppresses mid-range peaks, making the response curve much smoother.

If you are building a reflex system, keep the port free of damping material or any other obstructions, unless it is called for in the design.

One traditional rule of thumb is to put damping material on the interior of the back panel, one side, and either the top or bottom panel. The theory behind this advice is that one layer of damping material in each dimension will absorb rear reflections. In the real world one should aim for a margin of safety and put damping material on every interior wall except the speaker board. For final adjustment, use your ear. Add more material to any system that sounds loud at modest volume levels. Make sure the walls near the woofer are heavily covered and, if necessary, hang a curtain of material behind the woofer. If there is a single note boom, try stretching a layer of material over the back of the woofer, stapling it to the speaker board. Make sure this damping pad is stretched tightly.

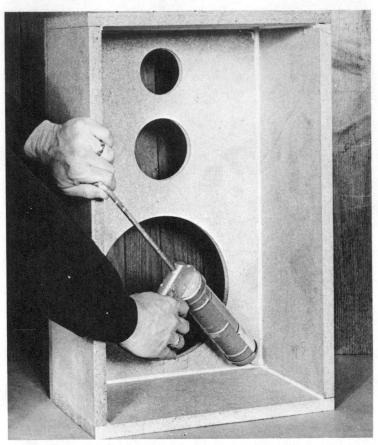

Fig. 3-6. Caulk all inner joints of speaker boxes. Project 2 shown here.

In addition to absorbing reflections, damping material can be used to increase the effective cubic volume of the box. It does this by changing the operation of the box air from adiabatic (constant heat) to isothermal (constant temperature). With isothermal operation the stuffing absorbs and gives up heat, maintaining a constant box temperature and reducing sound velocity. This shortens the wavelengths. In theory, you can add about 40% to box volume by stuffing; in practice, about 20% is the limit. To stuff a box, cut the damping material into small pieces and loosely fill the box. Don't force enough material into the box to compress it.

What kind of damping material? Fiberglass is probably the most widely used because it is widely available and its characteristics are well known. The acoustical grade comes in packages of one square yard of material 1″ thick. It is the preferred kind because it stays in place better than looser kinds. Ask for Radio Shack Cat. No. 42-1082. You can substitute other materials such as polyester batting, rug underlayment, or old rags; but beware of dense materials that would significantly change box volume.

●**Rule 4: Install Drivers From Outside The Box.** In earlier days the typical high fidelity speaker was bolted to the rear of the speaker board (Fig. 3-7). In some enclosures that is the only practical way to mount a driver; but, if the panel is very thick, such mounting produces a cavity at the front of the speaker that can color the sound. The sharp edge of the speaker cut-out may diffract the sound waves, producing interference which causes dips and peaks in the mid-range.

Front mounting solves those problems and permits a more vibration-proof box because there is no need for a removable panel. When every panel is glued to the others, rather than held by screws, it's unlikely that rattles can develop with use.

Some speakers come with a mounting gasket on the rear of the frame; others have no rear gasket. To install a speaker with no rear gasket, set the enclosure on its back and wire the speaker. Run a

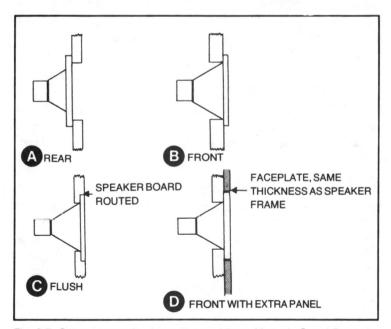

Fig. 3-7. Several methods of installing speakers. Methods C and D are the preferred ones, A is least desirable.

bead of silicone rubber compound around the edge of the speaker cut-out and press the speaker into place. Leave the box in that position for several hours, preferably overnight, until the silicone glue sets. The silicone rubber forms a perfect gasket and glues the speaker to the board. For speaker systems with very large woofers, 15" or larger, you can install T-nuts for speaker holding bolts. Drill speaker mounting holes one size larger than the bolts you intend to use, then drive the T-nuts into the holes on the back of the speaker board before you assemble the box. For example, if you want to use 3/16" mounting bolts, you would drill ¼" holes and drive 3/16" T-nuts into the holes at the rear of the board. With T-nuts in place, you can run a bead of silicone rubber around the cut-out, to make a gasket, but you won't have to leave the box on its back after installing the speaker.

●**Rule 5: Choose An Open Grille Cloth.** Don't cover super tweeters with a thick drapery material. Use a grille cloth made for the purpose, or test any cloth that you want to use by holding it up to the light. If you can see through it, it is probably a suitable choice.

A better test is to hang the cloth over your speakers and listen. If you can hear any change in the sound when the cloth goes on, try to find a better material. Obviously you should try to find a material that is both acoustically acceptable and has the kind of appearance you like.

FINISHING YOUR SPEAKERS

The kind of finish you put on your speaker cabinets depends on the material you use for the box. Particle board can be painted, covered with an adhesive-backed plastic veneer, or wood veneer. For plywood you will probably choose a more conventional finish, such as stain and varnish.

The final appearance of your cabinets depends on how well you fit the parts together and, perhaps even more, on how much time you spend with sandpaper. Start with a medium grit paper to remove uneven projections and to smooth poorly fitted trim or other parts. Then go to fine paper for the overall sanding. To check your work, adjust a bright light at a 30 degree angle to the work and observe the wood from all angles. Small scratches or saw marks that don't show on raw wood seem to leap to the surface after a stain is applied.

One of the easiest finishes to apply is an oiled wood finish. Stain the cabinet to the color you want, then coat it with a mixture of 3 parts linseed oil and 1 part turpentine. Make up the oil-turpentine mixture; then apply it with a rag soaked with the liquid mix. Allow it to stand for a half hour or more; then wipe it clean with a dry cloth. Sand lightly with a 360 grit wet or dry sandpaper and apply another coat. Do this until you have put 3 coats on the enclosure. Then rub it down. Repeat the process with a further coat once or twice a year for the first few years.

Another easy finish is paste shoe polish. Stain the enclosure, then add the polish for more color and a wax finish. Brown shoe polish over fruitwood stain gives a reddish, almost mahogany, color. If you want other effects, practice with multiple coats of various shades of shoe polish over stains on a piece of waste material that is identical to the wood used in the cabinet.

If you apply conventional finishes, such as varnish or lacquer, beware of incompatible combinations. The easiest way to make sure that your stains and final finishes will work well together is to buy a stain and varnish of a single brand name from a knowledge-able paint dealer. For a smooth looking finish, choose a satin or dull varnish, at least for the final coat. And don't use wax or polish on any cabinet until the final coat of varnish has been applied.

MORE BOX CONSTRUCTION HINTS

If you are making a nailed box, such as a particle board enclosure with butt joints, make sure you have a solid surface under your work while you drive in nails. A concrete floor is ideal. Even a slight vibration in your work surface while you are nailing makes tight joints virtually impossible to attain. Bent nails are a symptom of the problem.

You can fill minor gaps with latex caulking compound, but if the gaps are larger than about ¼", use a piece of wood or particle board. Fill the hole where the speaker cord goes into the box with latex or silicone rubber caulking compound. It's a good idea to knot the cord on the inside of the box so it won't pull through and damage the speakers or other components if someone trips over it. For convenient speaker connections, install a screw terminal strip in the back by drilling two holes, for each terminal, and bringing the speaker leads out through each hole and soldering them to the terminals. Then glue down the strip to the exterior surface of the back. Tacks will hold the strip while the glue sets. Fill the terminal holes on the inside of the panel with latex caulking compound. For speaker systems with crossover networks, the crossover compo-nents and screw terminals can be installed on a piece of ¼"

Table 3-1. Parts List for Project 2.

¾" Particle Board:

2	8⅞" × 22½"	Sides
2	8⅞" × 15½"	Top & bottom
2	15½" × 24"	Speaker board & back

⅜" Plywood:

1	15½" × 24"	Faceplate

Pine:

2	1" × 12" × 24½"	Slab sides

¼" Hardboard:

1	8" × 8"	Crossover board

Speakers & Components:

1	10" woofer	Radio Shack Cat. No. 40-1331B
1	4" mid-range	Radio Shack Cat. No. 40-1282
1	cone tweeter	Radio Shack Cat. No. 40-1270
2	8-ohm L-pads	Radio Shack Cat. No. 40-980
*1	15 μF capacitor	(supplied with mid-range speaker)
*1	4 μF capacitor	(supplied with tweeter)
*70 ft.	#20 magnet wire	For woofer low-pass filter
(½ lb. will make coils for 2 speakers)		

Miscellaneous:

Fiberglass insulation Radio Shack Cat. No. 42-1082 Grille cloth
*Or use Radio Shack's 3-way crossover network—Cat. No. 40-1299.

hardboard which can be glued to the interior of the back. A small "window" in the back allows access to terminals and control shafts.

Check your finished enclosures for quality by rapping each panel with your knuckles. Listen for a dead, relatively high-pitched sound that means solidity. If the box sounds like a drum, the panel isn't stiff enough. A rattle tells you something is loose. If the panels are solid and stiff enough to emit a rather high-pitched sound on your rapping, you can be assured that the enclosure is well made and that the damping material will take care of the residual panel resonance.

In planning the projects in this book several construction methods were chosen to illustrate the various ways of building a satisfactory box. You can substitute other construction procedures, but remember to plan for the recommended cubic volume.

PROJECT 2: DESIGNING AND BUILDING AN ENCLOSURE

Here we will design and build a closed box enclosure for a 3-way system using a 10" woofer, a 4" mid-range speaker, and a small cone tweeter (Table 3-1). Such a combination represents one of the most popular kinds of main speaker systems.

The woofer is an acoustic suspension model with a high compliance foam roll outer suspension. An appropriate cubic volume for this speaker is about 1.5 cubic feet—more on how to choose box volume for your speaker in the next chapter; for now you'll just have to accept that figure. Going back to Fig. 3-1, and checking the horizontal line for 2500 cubic inches (1.45 cu. ft.), we get these dimensions: 8½" × 14" × 23". Multiplying those values, the cubic volume will be about 2737 cubic inches, adequate to allow about 2500 cubic inches after subtracting 10% for internal space occupied by the speakers and a crossover network.

Now for some compromises. Any time there is a good reason to vary the ideal dimensions, do it. Here we can cut the internal length to 22½" and use 24" material for the basic box, an advantage because scraps are often cut into 24" widths or lengths. We will increase the depth to 8⅞" to maintain approximately the same internal volume. The new internal dimensions are: 8⅞" × 14" × 22½", giving a volume of about 2796 cubic inches.

For an easy-to-build enclosure, with no power tools needed, we will choose the design shown in Fig. 3-4, with butt joints. The

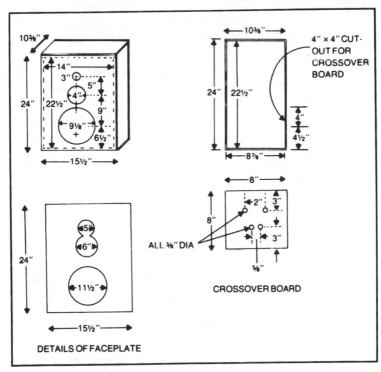

Fig. 3-8. Construction details for basic enclosure of Project 2. Pine slab sides are not shown here.

47

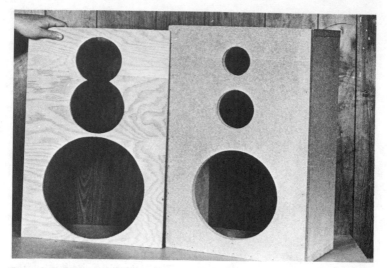

Fig. 3-9. Faceplate, left, is made like a grille board.

depth, 8⅞", plus 1½" for the speaker board and back, plus ⅜" for an external front panel to hold the grille cloth away from the front-mounted speakers just happens to come out right (10¾") so that we can use 1" × 12" pine for slab sides without ripping it. The actual width of 1" × 12" material is 11¼", which permits a ½" extension beyond the front board.

Cut out the parts according to the plans in Fig. 3-8, including the cut-outs in the speaker board. Begin assembly by gluing and nailing the shell, top, bottom, and sides together. Then glue and nail on the speaker board. It's a good idea to start the nails in each piece while it is supported by a concrete floor or other solid work surface, then place the piece onto the glued parts and nail it down. Wipe the excess glue from each joint; then caulk the internal joints with latex caulking compound.

Next, install the faceplate (Fig. 3-9) on the front of the basic box with glue and small nails (Fig. 3-10). Before installing the back panel you must mount the crossover network and the L-pads on the crossover board and glue the board inside the back. For easy wiring you can use a ready-made crossover network. The new Radio Shack 3-way crossover network (Cat. No. 40-1299) offers a choice of crossover frequencies. Read the instructions that come with the crossover network and choose the connections that provide 1600 and 7000 Hz crossover points. Or you can use the general purpose crossover network diagrammed in Fig. 3-11. This homemade crossover network makes use of the 15 μF capacitor supplied with the mid-range driver and the 4 μF capacitor supplied with the tweeter. All you have to add is a 1 mH choke. For instructions on making the choke, see Chapter 7.

To wire the general purpose crossover network follow this procedure:

- ●Prepare an 8″ × 8″ crossover board according to the plan in Fig. 3-8.
- ●Install push-button terminals (Radio Shack Cat. No. 274-621) from the smooth side of the board.
- ●Install L-pads (Radio Shack Cat. No. 40-980) from the rough side of the board.
- ●Install a 2-lug terminal strip on the rough side, as shown (Fig. 3-12).

Fig. 3-10. Glue and nail faceplate to front of enclosure.

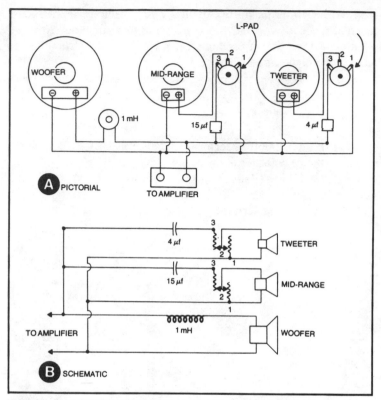

Fig. 3-11. Pictorial and schematic diagrams of the wiring circuit for Project 2.

● Run hook-up wire from lug #1 of left L-pad to lug #1 of right L-pad, then to right lug of push-button terminals or to the right lug of the 2-lug terminal strip which is connected to the right push-button lug. (Note: left or right as seen from the rough side of the board.)

● Remove the 4 μF capacitor from the tweeter and connect one lead to the left push-button terminal, one lead to pin #3 of the left L-pad.

● Connect one lead of 15 μF capacitor to left push-button terminal, other lead to pin #3 of right L-pad.

● Connect one lead of 1 mH coil to left lug of push-button terminal, other lead to left lug of terminal strip.

● Prepare 3 pieces of double conductor lamp cord about 18" long. Connect the woofer leads to the 2-lug terminal strip, the positive lead to the left lug. Connect the mid-range lead to terminals #1 and #2 of the right L-pad, and the tweeter leads to pins #1 and #2 of the left L-pad. In each case the positive lead goes to pin #2.

Note: The term left and right used above refer to the component positions as seen from the rough (interior) side of the crossover board. Check Figs. 3-12 and 3-13.

●Solder all connections with rosin core solder.

●Glue the crossover board over the cut-out in the rear panel, using latex caulking compound or silicone rubber sealer.

While the 1 mH coil is recommended, you can get by without it. Just make sure you use the capacitors in the mid-range and tweeter circuits.

Cover the interior of the sides, top and bottom, and back panel with a layer of fiberglass insulation. Then glue and nail down the back panel. Inspect the seam around the back for possible air leaks. If you find any questionable points, fill the gap with caulking compound on the outside. Then work through the woofer hole to push aside the damping material and caulk the inner seam.

It's a good idea to paint the front and top panels flat black, so the speakers won't show through the grille cloth.

Glue and nail on the pine slab sides next (Fig. 3-14). Sand, stain, and finish them before installing the speakers. To mount the speakers, set the enclosure on its back, wire the speakers, then,

Fig. 3-12. Wired crossover board. Upper wire goes to tweeter, right one to mid-range speaker, and lower one to woofer.

working with one at a time, run a bead of silicone rubber sealant around the cut-out in the speaker board. Press the speaker down into place and twist it slightly to make a good seal. After doing all the speakers, leave the cabinet in that position overnight so the silicone rubber will set. The silicone rubber will make a perfect seal and glue the speaker to the box.

Connect the speakers to your receiver with the power turned off. Turn the volume control down before turning on the receiver, then turn it up just enough to hear the speakers clearly. Check to see that the highs come from the tweeter and the lows from the woofer. Rotate the L-pad controls to see if they are working right. Then reduce the mid-range and tweeter level to a minimum. Bring up the mid-range level until the sound from that range blends with that of the woofer. Next adjust the tweeter until it blends. Try to get the system to sound as if only one speaker were working but at a level where you can hear all frequencies in proper balance.

When you are satisfied that everything is working right, install the grille cloth. Staple it under the bottom of the cabinet;

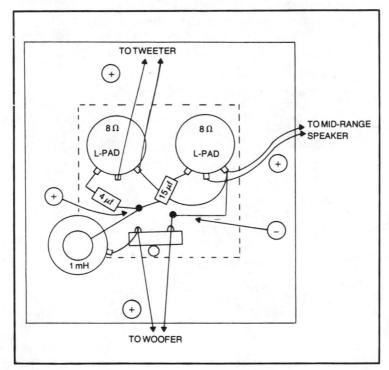

Fig. 3-13. Suggested layout of components for crossover network of Project 2.

Fig. 3-14. Pine slab sides complete the enclosure.

Fig. 3-15. Partially built and a finished enclosure with speaker components used in Project.

then stretch it up and across the top, stapling it again on the back. If there are any objectionable raw edges on the grille cloth, run a thin strip of wood, or a plain narrow ribbon to match the color of the cloth, along the edges. This completes the design and construction of the project (Fig. 3-15).

If you prefer, you can use external pieces of wood for the top and bottom and wrap the grille cloth around from side to side (Fig. 3-5). A more conventional enclosure could be made from birch plywood, using bevel joints. Some listeners might consider such a hardwood enclosure more attractive, but the sound would be no better than that of the enclosure described here.

4

Closed Box Speaker Systems

A closed box is the simplest speaker enclosure you can make, both in principle and in construction. One of its great virtues is that it forgives minor mistakes in box design. There is just one significant design problem to solve: what cubic volume is appropriate for the speaker.

CLOSED BOX BASICS

An unbaffled speaker's resonance is determined by the mass of the moving parts and the compliance of the suspension. When the speaker is installed in a closed box, the pressure of the trapped air in the box restricts cone movement to some extent. In effect, a speaker in a box is stiffer than in free air, and this stiffness raises the frequency of resonance. The smaller the box, the more it raises the resonance frequency (Fig. 4-1).

A certain volume of air itself has no special value of stiffness; that property depends on the size of the piston coupled to it. The air in a box resists the movement of a large piston more than that of a small one. Pressure is measured in force per unit area, such as pounds per square inch (lbs./sq. in.), so for a large speaker cone the pressure must be greater than for a small cone. And when the large cone moves, it changes the pressure against it much more than does a small cone. In fact, the compliance of box air varies inversely with the square of a driver's cone area. Since a 10" speaker has about 4 times the area of a 5" speaker, a given box volume would have 16 times as much compliance for the small speaker as the larger one. Or, stated in terms of stiffness, the box

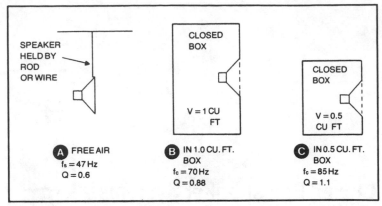

Fig. 4-1. How a speaker's frequency of resonance and Q varies when tested in free air and in boxes of different cubic volumes.

would be 16 times as stiff for the 10" speaker. If you are ever tempted to squeeze a large woofer into a box that is much smaller than is normally used for a speaker of its size, remember that relationship.

Putting a speaker into a small closed box not only raises the frequency of resonance, it also raises the Q (Fig. 4-1). If the speaker is a highly damped model, this may be desirable. Such a speaker may be overdamped if the box is too large. At the other end of the scale, a speaker with a magnet of minimum weight for the size of speaker can sound terrible if put into a box that is too small. That combination would produce an underdamped system that would boom loudly at resonance. These two factors, the final resonance frequency and speaker Q, make the choice of box volume an important one.

HOW TO FIT BOX VOLUME TO SPEAKER

Many audio fans buy a woofer, then choose a box volume for it. Sometimes the problem is reversed when the cubic space occupied by stereo speakers must be restricted to a certain pre-set figure. The problem in the second situation is to find the right speaker for a given box volume.

To get an idea of an appropriate box volume for each common speaker size, look at Fig. 4-2. This chart can be used as a rough design chart for any modern high compliance speaker that will be used in a closed box. Note that there is more allowance for speaker variation for large woofers than for small speakers. The upper line gives maximum cubic volumes; the shaded area shows a range of volumes that vary according to a speaker's compliance and size of magnet.

While Fig. 4-2 gives a set of box sizes that are right for most speakers, you may want to explore how variations in box volume would affect the bass range and response curve on your woofer. After studying Fig. 4-1, some speaker builders might be tempted to install the speaker described there in a larger box than those

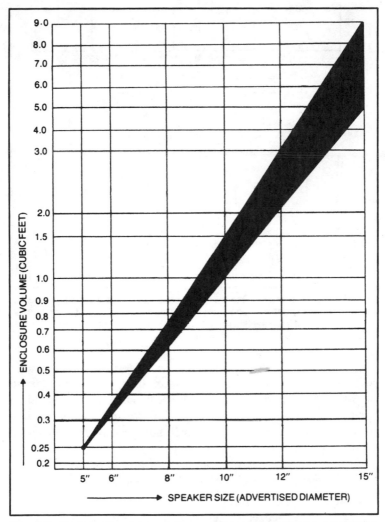

Fig. 4-2. Typical closed box optimum volumes for acoustic suspension speakers. A variation of 10% will have no significant effect on speaker performance.

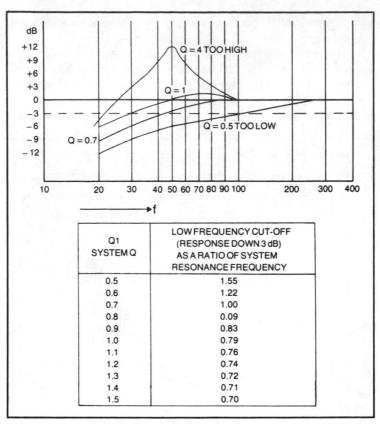

Q1 SYSTEM Q	LOW FREQUENCY CUT-OFF (RESPONSE DOWN 3 dB) AS A RATIO OF SYSTEM RESONANCE FREQUENCY
0.5	1.55
0.6	1.22
0.7	1.00
0.8	0.09
0.9	0.83
1.0	0.79
1.1	0.76
1.2	0.74
1.3	0.72
1.4	0.71
1.5	0.70

Fig. 4-3. Relation of system Q to low frequency response and cut-off. To find the cut-off frequency, get the ratio from the chart that is correct for system Q. Multiply that ratio by the system frequency, fc. Example: System has Qt of 0.9, fc = 70 Hz Cut-off f = 0.83 × 70 Hz = 58 Hz

shown in the belief that they would obtain a deeper bass response. A larger box might produce a weak bass response, due to overdamping, and the speaker's power handling ability would be reduced because the air in the larger box would have less control over the cone. A box can be too small, but it can also be too big for optimum performance.

To predict a speaker's bass performance in boxes of various cubic volume, three specifications must be considered. These are:

● f_s - the speaker's free air resonance frequency.
● Q_{TS} - the speaker's resonance magnification at f_s.
● V_{AS} - the speaker's compliance stated in terms of the air volume with an equivalent compliance for that speaker.

58

With these 3 values one can predict, for any suitable box volume, the following system characteristics:

●f_{CB} - frequency of the closed box speaker resonance, also called system resonance.

●Q_{CB} - the speaker's resonance magnification in the closed box.

●f_3 - the frequency where the bass response is down 3 dB, called the cut-off frequency.

To see how different values of Q affect the frequency response of closed box speakers, look at Fig. 4-3. As you can see from the graph, the higher the Q, the greater the hump in bass response near system resonance. The graph in Fig. 4-3 was based on the assumption that each curve represents a closed box speaker system with a system resonance at 50 Hz. If we are comparing curves for a single speaker, the frequency of resonance would be different for each value of Q, with low values of Q associated with a low resonance. That means that the low Q systems can yield a better bass response than is suggested by Fig. 4-3.

Most engineers aim for a Q_{CB} that is between 0.7, the value with the flattest response curve, and about 1 or slightly greater. Some rock music fans like a higher Q because it accentuates the bass, accepting speakers with Qs as high as 1.5 or even 2. Classical music buffs usually want a speaker with a Q_{CB} under 1. Knowing how much fullness you can accept in your speaker's bass response can be a useful guide to designing an enclosure for it. If you like prominent bass, try to get the Q_{CB} above 1; for refined, ultra clean bass, keep it at or below 1.

Here is an example of how to design a box for your speaker, assuming that you know your speaker's specifications. Suppose you have a 10″ woofer with the following specifications:

$$f_s = 30\,\text{Hz}$$
$$Q = 0.4$$
$$V_{AS} = 10\,\text{cu. ft.}$$

Suppose you want to accentuate the bass slightly, so you choose a final Q of about 1.2. The first step is to find the desired ratio of closed box Q (Q_{CB}) to the free air Q. This can be solved by:

$$Q_{CB}/Q = 1.2/0.4 = 3$$

This ratio will also be the ratio of closed box resonance to free air resonance, so the system will have a resonance frequency of about 90 Hz. But another important consideration is the bass cut-off frequency. Looking at Fig. 4-3, we find that with a Q of 1.2

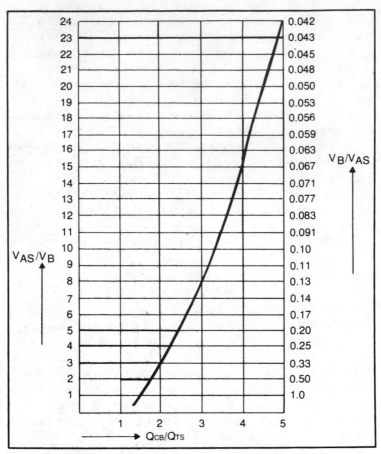

Fig. 4-4. Closed box design chart.

the bass cut-off occurs at 0.74 times the frequency of resonance, or, in this case, 0.74 × 90 Hz, or 67 Hz. If you want a more extended response, you could use a lower ratio of Q_{CB}/Q, which would also give a lower ratio of f_{CB}/f_s.

Assuming that 67 Hz is satisfactory, what box size should we choose to obtain a Q_{CB} of 1.2? To find box volume, we go to Fig. 4-4. There we find the desired Q_{CB}/Q ratio of 3 and move vertically to the curve, then across the graph horizontally to find the volume factors. The left margin of the chart tells us that we need a V_{AS}/V_B ratio of 8, or stated at the right margin in terms of V_B as a function of the speaker's V_{AS}, 0.13. To get the correct box volume, we can either divide the V_{AS} by 8 or multiply it by 0.13, whichever is more convenient. So:

$$V_B = 10 \text{ cu. ft.}/8 = 1.25 \text{ cu. ft.}$$

Here is a formula that permits you to check the values you get from Fig. 4-4:

$$A = (Q_{CB}/Q)^2 - 1$$
$$(\text{where A is } V_{AS}/V_B)$$

In the preceding example, where a Q_{CB} of 1.2 is desired:

$$A = (1.2/0.4)^2 - 1$$
$$= 3^2 - 1$$
$$= 8$$

If, for the same speaker, you decide you want to explore the kind of bass range you would get with a Q_{CB} of 0.7, the Q_{CB}/Q_{TS} ratio would be 1.75. Looking at Fig. 4-4, it appears that the V_{AS}/V_B ratio is about 2, so:

$$V_B = 10 \text{ cu. ft.}/2, \text{ or } 5 \text{ cu. ft.}$$

With a box volume of 5 cu. ft. the system resonance would be about 53 Hz, since the resonance ratio is about equal to the Q ratio, or 1.75 x 30 Hz. Going back to Fig. 4-3, we see that with a Q of 0.7 the bass cut-off occurs at the resonance frequency, so that would also be 53 Hz. For some kinds of music this would be a better choice, but if the speaker is to be driven hard, there is the question of power handling ability. With a relaxed air cushion behind the cone, the speaker can respond to lower frequency signals, but the cone will move farther, perhaps far enough to cause distortion or even damage at high power levels.

CLOSED BOX SUMMARY

The design procedure outlined above is based on a theoretical speaker with arbitrarily chosen parameters. In most acoustic suspension systems the V_{AS}/V_B ratio will be equal to 3 or greater. For small speakers, with an advertised diameter of 4" to 6", the ratio is often lower, sometimes as low as 1, which gives a Q ratio of 1.414. As you can see, optimum box volume can vary considerably according to the tastes of the builder. But remember these relationships:

- **Box Too Large** = Weak bass, reduced power handling.
- **Box Too Small** = Boomy bass, reduced bass range.

If you have a speaker with no specifications, you can use the more general chart in Fig. 4-2 as a guide to an appropriate box volume. Or, if you have some empty shipping cartons of various sizes, you can make some temporary test boxes and try your speaker with each one. Seal the cartons and cut a hole in one side large enough to match the cone area of your speaker. Hold your speaker over the holes in the cartons and listen to its performance with each one. If the bass sounds boomy, try a larger carton. Don't expect good sound with this kind of test; some bass response will be lost because of vibration in the carton walls. The thing to listen for is the frequency of the most prominent bass tones. If these are high enough to significantly accentuate male speech, try a larger box. For more precise tests, see the appendices. Before building a permanent enclosure, look over the construction rules in Chapter 3.

PROJECT 3: A CLOSED BOX SPEAKER SYSTEM

For this project we will choose an 8″ woofer with a polypropylene cone, Radio Shack Model No. 40-1021. The specifications for the woofer are:

$$f_s = 30 \text{ Hz}$$
$$Q = 0.41$$
$$V_{AS} = 2.55 \text{ cubic feet}$$

Assuming that a Q_{CB} of 1 is desired, the ratio of Q_{CB}/Q should be about 2.4. Going to Fig. 4-4 we find that this ratio can be obtained by choosing a V_B/V_{AS} ratio of about 0.2, so:

$$V_B = 0.2 \times 2.55 \text{ or } 0.51 \text{ cubic feet}$$

An enclosure volume of 0.51 cubic feet equals 881 cubic inches. In Fig. 3-1 we find the suggested internal dimensions: about 6⅜″ × 10″ × 16″. This allows some overvolume to make up for the space occupied by cleats and speaker components.

If the Q ratio is 2.4, as calculated, f_{CB} will equal 2.4 × 30 Hz, or about 72 Hz. With a Q_{CB} of 1 the bass cut-off will be (from Fig. 4-3) 0.79 × 72, or 57 Hz. Note that the term "cut-off" doesn't mean that there is no audible base below that frequency; it means that the bass response will be down 3 dB there. Closed box systems give a smooth and gradual roll-off in bass below the frequency defined by custom as the cut-off point.

Crossover Network

The small box makes a two-way system the practical choice (Fig. 4-5). By choosing Radio Shack polypropylene tweeter, Model No. 40-1374, you can build an all polypropylene system (Table 4-1). The suggested low-frequency limit for this tweeter is 3000-

Fig. 4-5. Project 3, complete except for grille cloth.

Table 4-1. Parts List for Project 3.

¾" Plywood:
2   7⅞" × 17½"   Sides
2   7⅞" × 11½"   Top & bottom
¾" Plywood or Particle Board:
2   10" × 16"   Speaker board and back
Pine:
4   ¾" × ¾" × 10"   ⎫
4   ¾" × ¾" × 14½"   ⎬   Cleats
4   ¾" × ¾" × 4⅞"   ⎭   Corner glue blocks
Grille Frame:
As desired
Speakers and Components:
1   8" Polypropylene woofer   Radio Shack Cat. No. 40-1021
1   1" Polypropylene tweeter   Radio Shack Cat. No. 40-1374
1   2-way crossover network   Radio Shack Cat. No. 40-1296
1   8-ohm L-pad   Radio Shack Cat. No. 40-977
1   Terminal plate   Radio Shack Cat. No. 40-625
Miscellaneous:
Fiberglass insulation   Radio Shack Cat. No. 40-1082
Grille cloth   As desired

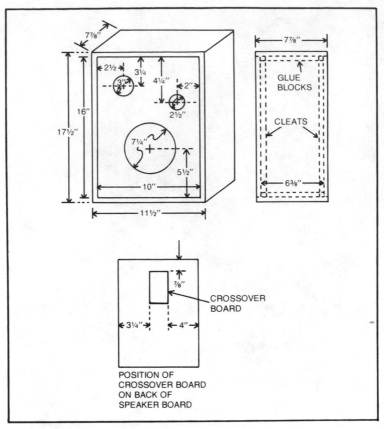

Fig. 4-6. Enclosure plans for Project 3.

Hz, the same as the upper-frequency limit for the woofer. To use the tweeter with a 3000-Hz crossover, a 12-dB/octave filter is desirable. Radio Shack's two-way crossover network is a 6-dB/octave type, and the crossover frequencies are not ideal for this combination of drivers. But it is possible to use this crossover network as a 3000-Hz sharp cut-off filter by an unconventional hook-up (Fig. 4-7). Note that this circuit puts a capacitor in series with the tweeter and a coil across it. Remember to remove the capacitor that is supplied with the tweeter before wiring the system.

Construction

This project uses ¾″ material, such as particle board or a good grade of plywood (Fig. 4-6). Assemble box sides with wood glue.

Then install cleats and corner glue blocks with wood glue. Caulk all joints. Install the back with silicone rubber sealant. Cover all interior walls with a 1″ layer of acoustical fiberglass. Install the crossover network on the back of the speaker board with a blob of silicone rubber sealant under it, top and bottom. After the rubber is set, invert the board and install the speakers with silicone rubber and install the L-pad with silicone rubber and the screws supplied with it. Wire the speakers. Then install the speaker board with silicone rubber sealant or latex caulking compound. Reverse the speaker board on your second enclosure so the stereo pair will be mirror images of each other.

Project Conclusion

The final tests on this project showed that f_{CB} was 74 Hz and Q_{CB} was about 0.96, very close to the projected figures. The system performed as anticipated, with smooth sound throughout the full frequency range.

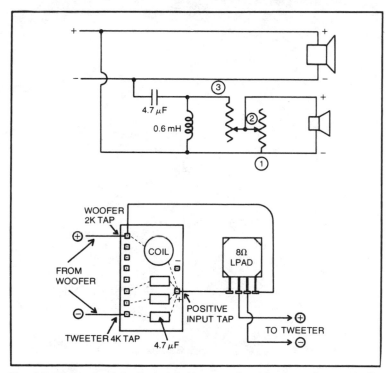

Fig. 4-7. Wiring circuit for Project 3. Unconventional hook-up gives a 12 dB/octave filter for tweeter.

PROJECT 4: AN ULTRA-COMPACT TWO-WAY SPEAKER

If your space for speakers is severely limited, this may be the project for you (Fig. 4-8). This 4″ woofer, with a Q of 0.35 and a V_{AS} of 0.23 cubic feet, requires a closed box volume of about 0.08 cubic feet. That translates into 138 cubic inches, a figure below the range of Fig. 3-1. Reference to a handy calculator gives the internal dimension of 3½″ × 5″ × 8″.

For the tweeter we will choose a ¾″ polycarbonate hard dome model, Radio Shack Model No. 40-1376. This tweeter has excellent dispersion characteristics. It comes with a capacitor high-pass filter

Fig. 4-8. Project 4 and Project 7. The speakers are the same, but the ported system of Project 7 requires a larger box.

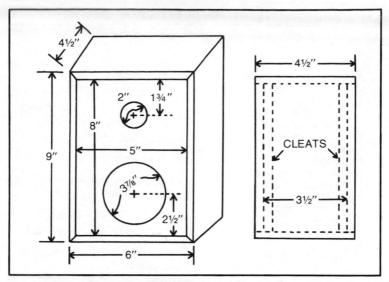

Fig. 4-9. Enclosure plans for Project 4.

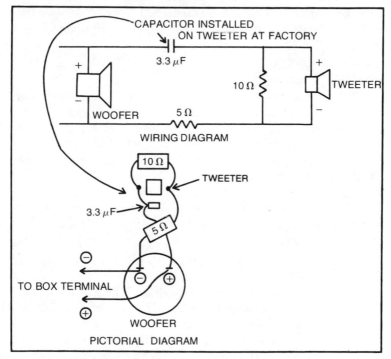

Fig. 4-10. Wiring circuit for Project 4.

Table 4-2. Parts List for Project 4.

½" **Plywood**
2	4½" × 9"	Sides
2	4½" × 6"	Top & bottom
2	5" × 8"	Speaker board & back

½" × ½" **Wood Strips:**
4	½" × ½" × 5"	⎫
4	½" × ½" × 7"	⎬ Cleats
4	½" × ½" × 2½"	⎭ Corner glue blocks

Grille Frame:
As desired

Speakers and Components:
1	4" Woofer	Radio Shack Cat. No. 40-1022
1	¾" Hard dome tweeter	Radio Shack Cat. No. 40-1376
1	5 ohm, 5 watt resistor	
1	10 ohm, 5 watt resistor	
1	Terminal plate	Radio Shack Cat. No. 274-625

Miscellaneous:
Fiberglass insulation	Radio Shack Cat. No. 42-1082
Grille cloth	As desired

wired on, so no crossover network is needed. Serving to balance the output of the tweeter to that of the woofer (Fig. 4-10) are two 5-watt resistors.

Construction

You can build this mini-enclosure from ½" plywood (Fig. 4-9). Construction procedure is much like that of Project 3 except for the speaker wiring (Fig. 4-10). Cover the inside surface of the box, except for the speaker board, with a 1" layer of fiberglass.

Install the speakers with a silicone rubber gasket under each, using #4 × ½" sheet metal screws. To shorten the connections between woofer and tweeter, install the woofer with its solder lugs up. With the screws holding the speakers in place, you can immediately proceed to the wiring. Glue the speaker board in place with silicone rubber sealant.

Project Conclusion

The final test showed the f_{CB} to be about 90 Hz with a Q_{CB} under 0.6. We would consider a larger system to be overdamped with such a low Q, but when the frequency of resonance is close to the range of male speech, such high damping is useful. You can't expect to get ultra low-frequencies (low bass) from such a small system, but the sound is clean.

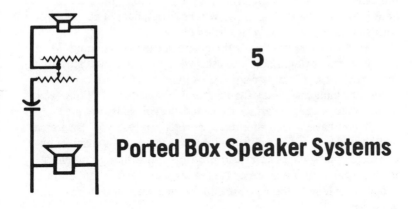

5

Ported Box Speaker Systems

As mentioned in Chapter 2, a ported box can be tuned by altering the area and length of the port. The problem is to know at what frequency to tune the box for best performance. Another design problem, similar to the one posed by closed box speakers, is how big to make the box. But first, let's consider some common types of ported systems.

KINDS OF PORTED ENCLOSURES

There are 3 main kinds of ported box systems: those with a simple hole in the speaker board for a port, those with a tube, or duct, behind the port, and those with a suspended diaphragm, or passive radiator, for a port substitute (Fig. 5-1).

The large enclosure with a simple port of large area is now almost extinct. One reason, we know now that optimum tuning frequency for many speakers is below 50 Hz, and to tune a box to any frequency below 50 Hz without a duct requires a tiny port, which can produce unmusical noises, or a tremendous enclosure, which few homes can tolerate. For example, if you wanted to tune a box to 30 Hz and if you insisted on a minimum port area of 7 square inches, equal to a round port of 3″ diameter, the cubic volume of the box would have to be 7 cubic feet.

The most popular method of tuning an enclosure is by a duct which increases the mass of vibrating air and permits lower frequency tuning. One precaution when using a duct is to select a port area that makes the duct shorter than the box depth by 2 or 3 inches. The larger the port area, the longer the duct must be for a given tuned frequency. One must reach a suitable compromise between a minimum port area and a maximum duct length. If an indicated duct length is too great for the box depth, the duct can be bent into an L shape (Fig. 5-2). Where such a duct is needed, you can use plastic drainpipe with a fitted elbow.

Another consideration is the shape of the port opening. Most tuning charts, including the ones shown in this book, are calculated for ports that are round or square. A long, narrow slot can have a different tuning frequency than a round or square port of the same area, so such extreme variations should be avoided if possible.

If you have a way of cutting perfect circles, you can use a paperboard tube for a port. Such tubes are used by rug manufacturers for shipping cores. You can probably get a tube at no charge from a local furniture store or carpet installer. Cut the tube to the proper duct length, then glue it into the speaker board so it is flush at the front.

Another way to install a duct is to make a rectangular one from ½″ material, such as those illustrated in Project 6. For this kind of duct it is easier to make the hole in the speaker board equal to the internal dimensions of the duct and butt the end of the duct against

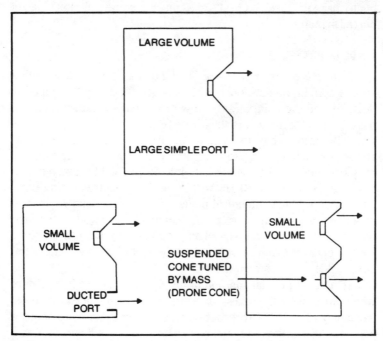

Fig. 5-1. Kinds of ported enclosures with action at resonance.

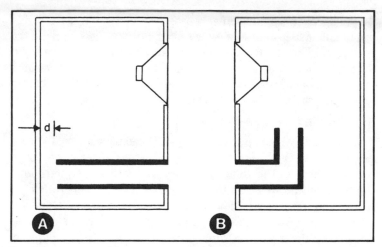

Fig. 5-2. How to install a long duct. A, Wrong if d is less than 3″. B, Right. Effective length of this duct is distance through center.

the speaker board. This means that you must subtract the thickness of the speaker board from the required duct length to get the proper length of the add-on duct. Install this kind of duct with screws or nails and glue.

Make sure the port is free of any damping material or other obstruction. A port or duct with a smooth inner surface is preferable so that air can move through it easily. This rule should be violated only in problem cases where stuffing the duct makes an audible improvement in the sound. Theoretically it is always bad because a stuffed port can't damp the driver as effectively.

The position of the duct is relatively unimportant if it is located at least 3 inches or more from the woofer. A port that is too close to the woofer will be more susceptible to reflected sound in the upper bass or mid-range. The use of damping material plus adequate distance between woofer and port can reduce this possibility.

The third kind of ported box, with a passive radiator instead of an empty hole, is useful for tuning a small enclosure to a very low frequency. The extra radiator, which is like a speaker cone with no magnet and no voice coil, can be tuned to very low frequencies because the diaphragm has more mass than air in a simple port. For easy tuning, some radiators have a center bolt to which cardboard discs can be bolted. Another method is to glue weights, such as steel washers, to the drone cone. A passive radiator should have high compliance and an effective cone area twice that of the woofer.

An old speaker can be used as a passive radiator if it has a highly compliant suspension. Just knock off the magnet and remove the pole piece. To tune it, use modeling clay as a temporary weight; then use silicone rubber sealer to glue an equal weight of washers or rings of solder to the cone. Place these masses near the center of the diaphragm in a balanced pattern.

HOW TUNING AFFECTS RESPONSE

A properly tuned ported box system can have a very flat bass response curve right down to cut-off (Fig. 5-3). If the box is tuned too high, there will be a hump in the frequency response; if too low, the bass will be weak. In the days of the classic bass reflex the instructions for builders of ported boxes always stated that the box should be tuned to the free air resonance of the speaker. Now we know that the optimum tuning frequency varies for different speakers from frequencies much lower than the free air resonance to frequencies above the free air resonance. It all depends on the Q of the speaker.

For example, a speaker with a high Q will peak at resonance, while a speaker with a low Q will have reduced response at its frequency of resonance. To prevent boom in a high Q speaker, a ported box should be tuned to a frequency well below resonance. But a speaker with a low Q can be overdamped at resonance unless the box is tuned to a higher than normal frequency.

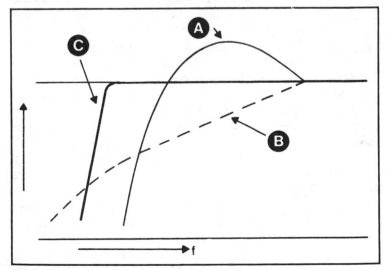

Fig. 5-3. Frequency response curves of bass reflex speaker when enclosure is tuned to a frequency that is (A) too high (B) too low and (C) optimum.

HOW BOX VOLUME AFFECTS RESPONSE

The effect of box volume on a ported system is much the same as for a closed box speaker: small boxes raise the Q and increase boominess more than do large boxes. This indicates a large box for speakers with a high Q. If a small box must be used, then the speaker should have a high Q with high damping.

Such terms as high and low Q or large and small boxes don't have much meaning without numbers attached to them. For ported box speakers, we can consider speakers with a Q of below 0.38 to be low and those with Qs significantly above that point to be high. Speakers designed specifically for use in ported boxes have significantly lower Qs than those designed for closed box use.

Whether a box is large or small for a given speaker depends on that speaker's compliance, or V_{AS}. A box that seems large in your living room may be acoustically small for a large woofer. A high compliance 15″ driver, for example, can have a V_{AS} of from 20 to 30 cu. ft., so a 6 cu. ft. floor model speaker cabinet, which would seem extremely large to most families, would be acoustically small for such a woofer. The same floor cabinet would be acoustically very large for an 8″ driver with a V_{AS} of 1 cu. ft. The ranks of high fidelity speakers include drivers with specifications that range from one extreme to the other and beyond. These specifications must be considered if you intend to design a ported box for your speaker.

THIELE PORTED BOX DESIGN

A. N. Thiele, an Australian engineer, has brought a high degree of order to the formerly haphazard task of designing a ported box for your speaker. To use the Thiele method, 3 kinds of driver specifications must be considered:

●f_s - the driver's frequency of free air resonance.
●$\dot{Q}_{TS}$ - the driver's resonance magnification at f_s.
●V_{AS} - The driver's compliance, stated in terms of the air volume that has an equivalent compliance for that driver.

Using these 3 values we can determine the 3 critical aspects of a ported box. These are:

1. Box volume (V_B).
2. Box resonance frequency (f_B).
3. System cut-off frequency (f_3).

Here we will follow a simplified design process, using charts to obtain factors which, when multiplied by driver specifications, give optimum values (1 and 2). If you want to pursue the design process further, you can use the pocket calculator design method shown in Appendix D. Or, for an even more sophisticated approach, there is the computer program, also in Appendix D. If you have access to a Radio Shack TRS-80 computer, you can quickly design an optimum enclosure for your speaker or readily explore how changes in box volume will change the frequency response curve. With the TRS-80 you can even obtain a predicted response curve for each box volume and tuning condition.

For now assume you want to do the job with nothing more than pencil and paper. Suppose you want to design an enclosure for an 8″ polypropylene woofer, Radio Shack Model No. 40-1021, mentioned in Chapter 4. Here, again, you must consider the parameters of the woofer:

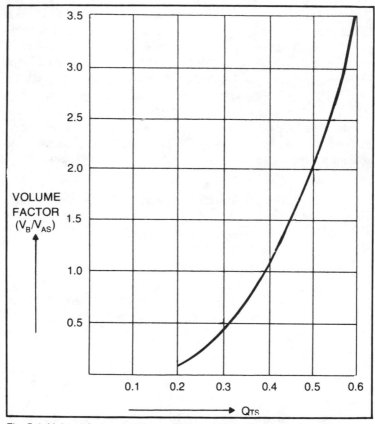

Fig. 5-4. Volume factor chart for ported box design. Find the point on the curve that corresponds to the Q of your speaker, then move to left margin to get volume factor. Multiply your speaker's V_{AS} by the volume factor to get the optimum ported box volume.

$$f_s = 30 \text{ Hz}$$
$$Q = 0.41$$
$$V_{AS} = 2.55 \text{ cubic feet}$$

The first step is to find the optimum cubic volume for the enclosure. Going to Fig. 5-4 and following a vertical line upward from 0.41 to the curve, then to the left margin, it appears that the volume factor is equal to about 1.15. Next, multiply 1.15 times 2.55 cu. ft., the woofer's V_{AS}, to get 2.93 cu. ft.

To get the correct tuning frequency (f_B) go to Fig. 5-5. Again locate the point on the curve directly above the Q value (0.41) and move to the left margin to get the tuning factor, which appears to be about 0.93. Multiplying 0.93 times 30 Hz (f_s), f_B should be 28 Hz. This completes the basic design, except for the dimensions of the box and port, but it would be helpful to know the bass cut-off frequency, f_3. For that you locate the V_B/V_{AS} ratio, 1.15 in this case, along the base of Fig. 5-6. Then move up to the curve and left to the margin to get the bass range factor. In this case, it appears to be about 0.9. The cut-off frequency will be about 0.9 times 30 Hz (f_s), or 27 Hz.

The charts in Figs. 5-4 and 5-5 are designed to give the optimum values for your woofer. If you want to build an enclosure that is smaller than optimum, you can estimate how much it would limit the bass range by referring to Fig. 5-6. For details on tuning frequency and response variations of such alternatives, use the calculator or computer programs in Appendix D.

Getting the Box and Port Dimensions

You can find the internal dimensions of the box from Fig. 3-1. That chart shows that the dimensions for 5000 cu. in., or 2.9 cu. ft., are approximately 11″ × 18″ × 29″, if each value is rounded off to the next higher whole number. For more exact work, use a calculator, but such accuracy is unnecessary. These dimensions will allow a bit more than the usual 10% overvolume, a good practice to follow with ported boxes. Because of unpredictable internal losses, ported systems often need a bit extra volume to reach the performance level suggested by theory.

The final step in box design is to find the correct port area and length to tune the box to 28 Hz. To do that we consult the port design charts in Figs. 5-7 to 5-11. The goal is to find the longest port that will tune the box correctly and yet fit within the 11″ box depth with a few inches to spare. We begin with Fig. 5-7, the duct

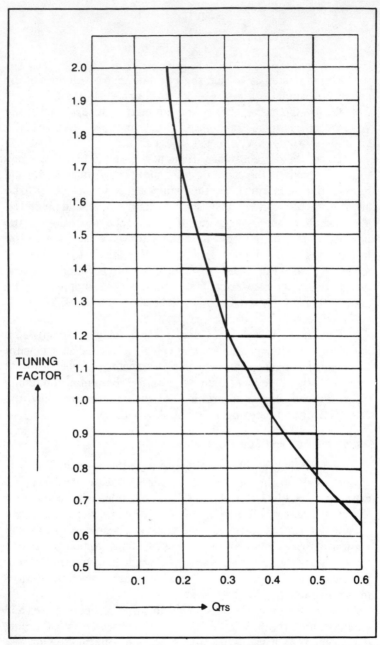

Fig. 5-5. Tuning factor chart. First get the tuning factor that matches the Q of your speaker. Then multiply the frequency of free air resonance by the tuning factor to get the correct tuned box frequency (f_B).

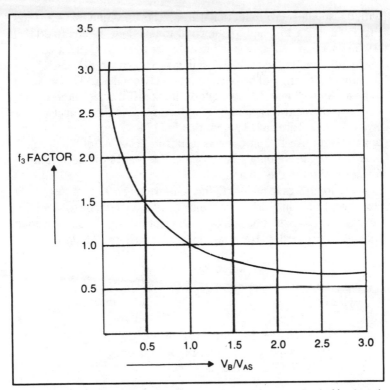

Fig. 5-6. Bass range factor chart. Divide box volume by speaker's V_{AS} to get V_B/V_{AS}. Find f_3 factor from chart. Multiply f_3 factor by speaker's free air resonance frequency to find the bass cut-off frequency of ported box system.

VOL. (FT.³)	20	25	30	35	40	45	50	60	70	80	90	100
						FREQUENCY (Hz) LENGTH IN INCHES						
0.5						7	5⅜	3¼	2	1¼		
0.75					5⅝	4	3	1⅝	⅞			
1				5½	3⅞	2¾	1⅞	⅞				
1.25			6	4	2¾	1⅞	1¼					
1.5		7⅝	4¾	3⅛	2	1⅜	¾					
1.75		6¼	4	2½	1½	1						
2		5⅜	3¼	2	1¼							
2.5	7	4	2¼	1¼								
3	5⅝	3	1⅝	⅞								
3.5	4⅝	2½	1¼									
4	3⅞	2	1									
5	2¾											

Fig. 5-7. Duct length for port with 3.14 in.² of area (2 in. tube).

with the smallest cross-sectional area. There we find that a 2″ tube that is between 1⅝″ and 3″ long will tune a 3-cu. ft. box to 28 Hz, so the duct is needlessly small in area. At the next chart, Fig. 5-8, the length of a 3″ tube for 25 Hz tuning is 8″, and for 30 Hz, 4¾″; so the correct length will be between these two values. This appears to be a useful choice, but to explore the possibility of a larger port, we go to the next chart, Fig. 5-9. Here we find that a duct with a cross-sectional area of 14 sq. in. that is 11″ long will tune a 3-cu. ft. box to 30 Hz. An 11″ duct, measured from the front of the speaker board, would leave only a ¾″ space at the rear of the duct. The 3″ diameter tube is the obvious choice.

Figure 5-9 gave no length figure for tuning a 3-cu. ft. box to 25 Hz with a 14-sq. in. duct. This indicates that it would be too long for most 3-cu. ft. boxes, but suppose you want to find out just how long such a duct must be. Here is a formula that lets you do that:

$$\Delta L_V = - (\Delta f_B) (2L_V/f_B)$$

where:

ΔL_V = change in vent length in inches
Δf_B = required change in box frequency
L_V = length of original duct

Here, to find the necessary change of length:

$$
\begin{aligned}
L_V &= -(-2 \text{ Hz}) (2 \times 11″/30 \text{ Hz}) \\
&= (2) (0.37″) \\
&= 0.74″
\end{aligned}
$$

This indicates that we should add about ¾″ to the length of the duct, making it 11¾″ long. Such a duct would have to be bent (Fig. 5-2).

Notice that this formula assigns a negative value to the change, showing how much duct to remove when the frequency is too low. For cases where the frequency is too high, as in this one, the change in frequency must be a minus number. The two minuses cancel, giving a positive number that is equal to the number of inches to add.

FINE TUNING BY EAR

The design process described above is based on Thiele data; however if you happen to find a chart of Thiele's original

VOL. (FT.³)	FREQUENCY (Hz) LENGTH IN INCHES 20	25	30	35	40	45	50	60	70	80	90	100
0.5								8⅜	5⅜	3¾	2½	1⅝
0.75		USE SMALLER PORT				10¼	7⅞	4¾	3	1¾	⅞	
1					9⅝	7⅛	5⅜	3	1⅝	¾		
1.25					7¼	5¼	3⅞	2	⅞			
1.5				8⅛	5¾	4	2⅞	1¼				
1.75				6⅝	4½	3⅛	2⅛	¾				
2			8⅜	5½	3¾	2½	1⅝					
2.5			6¼	4	2½	1½	¾					
3		8	4¾	3	1¾	⅞						
3.5		6½	3¾	2¼	1¼							
4	9⅝	5⅜	3	1⅝	¾					USE LARGER PORT		
5	7¼	3⅞	2	⅞								
6	5¾	2⅞	1¼									
7	4½	2⅛	¾									
8	3¾	1⅝										
10	2¼	⅞										
12	1¾											

Fig. 5-8. Duct length for port with 7 sq. in. area (3 in. tube).

VOL. (FT.³)	FREQUENCY (Hz) LENGTH IN INCHES 20	25	30	35	40	45	50	60	70	80	90	100
0.5										8¾	6¼	4½
0.75									7¼	4¾	3⅛	1⅛
1								7⅜	4⅝	2⅛	1½	
1.25		USE SMALLER PORT					9	5⅜	3	1⅝		
1.5						9⅜	7	3⅞	2	¾		
1.75					10⅜	7⅝	5½	2⅞	1⅜			
2					8¾	6¼	4½	2⅛	¾			
2.5				9¼	6⅜	4⅜	3	1⅛				
3			11	7¼	4¾	3⅛	2					
3.5			8⅞	5⅝	3⅝	2½	1¼					
4		12	7⅜	4⅝	2⅞	1½						
5		9	5¼	3	1⅝							
6		7	3⅞	2	¾							
7	10⅜	5½	2⅞	1⅜							USE LARGER PORT	
8	8¾	4½	2⅛	¾								
10	6⅜	3	1									
12	4¾	1⅞										

Fig. 5-9. Duct length for port with a 14 sq. in. area (3¾ in. x 3¾ in.)

FREQUENCY (Hz)
LENGTH IN INCHES

VOL. (FT.³)	20	25	30	35	40	45	50	60	70	80	90	100
0.5												
0.75												9⅜
1											7	2⅝
1.25										10	4¼	1¼
1.5									9⅝	6⅜	2½	
1.75								8⅜	6⅞	4¼	1⅜	
2							9⅜	6⅜	5	2⅞		
2.5						9¼	6⅝	5⅛	3¾	1⅞		
3				11⅝	9⅞	7	4⅞	3⅜	2¾	1⅛		
3.5			10⅞	9⅝	7⅞	5⅜	3⅝	2⅛	1⅜			
4			8⅜	6⅞	6⅜	4¼	2⅝	1¼				
5			6⅝	5	4⅜	2½	1¼					
6			5¼	3¾	2⅞	1⅜						
7		11¼	3⅜	2¾	1⅞							
8		9⅜	2	1⅜	1⅛							
10	12¾	6¾										
12	10	4⅞										

USE SMALLER PORT

USE LARGER PORT

Fig. 5-10. Duct length for port with a 25 sq. in. area (5 in. x 5 in.).

81

FREQUENCY (Hz)
LENGTH IN INCHES

VOL. (FT.³)	20	25	30	35	40	45	50	60	70	80	90	100
0.5												
0.75												
1										10¾	10½	11⅞
1.25										8	7¼	7⅜
1.5										6	5	4¾
1.75										4½	3½	3
2									9⅝	2½	2⅜	1¾
2.5								8⅞	7¾	1	¾	¾
3							9⅜	6½	5			
3.5								4¾	3¼			
4						10½	7½	3⅜	1⅞			
5					10¾	7¼	4¾	1⅝	1			
6					8	5	3					
7				9⅝	6	3½	1¾					
8				7⅝	4½	2⅜	¾					
10		11⅞	8⅞	5	2½	¾						
12			6½	3¼	1⅛							

USE SMALLER PORT

USE LARGER PORT

Fig. 5-11. Duct length for port with a 49 sq. in. area (7 in. x 7 in.).

82

alignments, you will note that the cubic volumes do not match those obtained by the simplified design method given here. Thiele's data is based on pure theory, but real systems have losses that prevent perfect agreement between predicted response and measured performance. The cubic volume obtained by charts in this book allow for about 30% overvolume, as specified by Richard Small, D. B. Keele, and others, to make up for typical losses. Such losses will vary according to the quality of the box, location of damping material, and even the driver itself. In practice, the 30% overvolume seems to work well.

If you build a system and find that the bass is slightly weak or, on the other hand, is boomy, you can make the final port adjustments by ear. For example, if the bass is weak, you can tune the box to a higher frequency by shortening the tuning duct. Then listen again. If the bass is boomy, try lengthening the duct, or if that isn't possible, use a duct with a smaller cross-sectional area (Fig. 5-12). You may not get a system that is theoretically perfect by such experimenting, but you can often improve one that looks perfect on paper. After all, you are building a speaker system to listen to, not one to fit some formula.

PROJECT 5: A PORTED BOX SYSTEM

Here is a ported box project based on the enclosure design that was completed in the preceding pages. In addition to the 8″ poly-

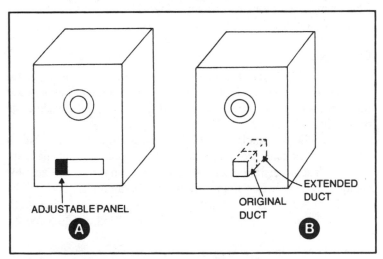

Fig. 5-12. How to lower the frequency of tuning of a bass reflex enclosure (A) Simple port, reduce port area (B) Duct port, increase length of duct.

Fig. 5-13. Project 5, complete except for grille cloth.

propylene woofer, it uses a ferrofluid cooled 4" mid-range speaker and an ultra wide range leaf tweeter (Fig. 5-13). The plans for the enclosure are shown in Fig. 5-14, and the parts list is in Table 5-1. The wiring circuit for this project is identical to that shown in Fig. 7-12 in Chapter 7.

Construction

Although ¾" material is used for the enclosure, some extra bracing is desirable. A single 1" × 4" vertical brace is adequate for the back panel. Glue and screw, or nail, the brace to the back, edge on and just off center.

Assemble the shell. Then install cleats to serve as air stops against the back and front panels. Caulk all joints except the front panel; if instead installed with screws, you can remove it later for changes in tuning. Use foam weather stripping on the back of the speaker board to insure an airtight seal. Apply the weather stripping just inside the screw line so the screws will not displace it.

Project Conclusion

This system gives excellent performance, particularly with

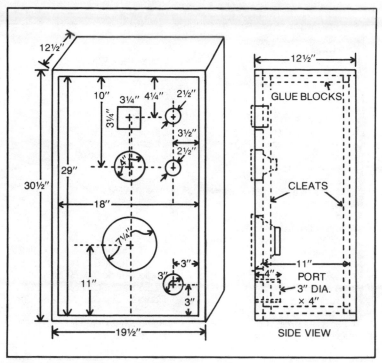

Fig. 5-14. Enclosure plans for Project 5.

receivers or amplifiers that have a suitable infrasonic filter to eliminate turntable rumble or other low frequency noise.

Compared to the closed box system with the same woofer, Project 3 in Chapter 4, this ported speaker offers extended bass range. And the 3-way system, with a control on the mid-range driver as well as the tweeter, permits more careful tailoring of system response. The differences are subtle for some ears, but well worth the extra expense and effort for a serious listener.

PROJECT 6: A DOUBLE CHAMBER REFLEX

Here is an enclosure that enables the 8″ driver to give remarkable bass performance. It was inspired by a design by audio engineer George Augspurger, who borrowed the idea from English experimenters.

Unlike a normal ported box, this one has two chambers. Each chamber is ported, and there is a connecting port in the partition (Fig. 5-16). All three ports have the same dimensions, but the primary chamber, that holds the woofer, has just twice the cubic

Table 5-1. Parts List for Project 5.

¾″ Plywood:

2	12½″ × 30½″	Sides
2	12½″ × 19½″	Top & bottom

¾″ Plywood or Particle Board:

2	18″ × 29″	Speaker board and back

Pine:

1	1″ × 4″ × 27″	Brace for back
4	¾″ × ¾″ × 18″	Cleats
4	¾″ × ¾″ × 27½″	Cleats
4	¾″ × ¾″ × 9½″	Corner glue blocks

Paperboard Tube:

1	3″ Inside diameter × 4″	Port

Or make a port with square cross section 2⅝″ × 2⅝″ I.D.

Grille Frame:
As desired

Speakers and Components:

1	8″ Polypropylene woofer	Radio Shack Cat. No. 40-1021
1	4″ Mid-range driver	Radio Shack Cat. No. 40-1282
1	Leaf Tweeter	Radio Shack Cat. No. 40-1375
1	3-way Crossover network	Radio Shack Cat. No. 40-1299
2	8 ohm L-pads	Radio Shack Cat. No. 40-977
1	Terminal plate	Radio Shack Cat. No. 274-625

Miscellaneous:

1	Pkg. Fiberglass	Radio Shack Cat. No. 42-1082
	Grille cloth	As desired

volume of the secondary chamber. This arrangement allows the enclosure to resonate at two frequencies instead of one. The primary chamber is tuned to about 70 Hz, but at lower frequencies the two chambers act as one large chamber, so the system is also tuned to about 35 Hz. In the enclosure pictured, the lower frequency turned out to be about 32 Hz. This double tuning provides bass reinforcement over a much wider band than would be possible with a single chamber. Moreover, the staggered resonances offer useful speaker damping over a wider band with reduced bass distortion in that range. The soft dome mid-range driver and tweeter complete this unusual system.

Construction

Before cutting out the parts, alert yourself to the asymmetrical layout of the speaker board (Fig. 5-16). To use this arrangement in a stereo pair you should reverse the plan for the second speaker board so that the two speakers will be mirror images of each other. This can easily be done by inverting one speaker board before gluing on the ducts. Note that you should also reverse the position

Fig. 5-15. Double chamber enclosure requires 3 ducted ports.

of the duct on the partition so that the duct is behind the speakers rather than the front ducts.

The front and back panels are glued to the particle board liner and to the partition rather than screwed in because no tuning changes are required with this enclosure. You can add damping to the woofer, if desired, by gluing small pieces of fiberglass to the back of the woofer frame.

Assemble the enclosure shell, then install the particle board liner. Note that the liner pieces should be 11″ wide (Table 5-2), to provide the correct interior depth, and it should be installed to leave a 1¼″ space at the front edge and a ¾″ space at the rear for the front and back panels. Install the liner with glue and 1″ nails or screws.

Next, cut a partition from ¾" particle board. Make this partition 11" wide and to the precise length that will permit it to fit snugly across the box, making contact with the liner at each side. Cut two pine cleats from ¾" × ¾" material and install them with glue and 1¼" screws, located so their upper edges are 101/16" from the liner under the top. Note that this location will make the upper chamber depth about 95/16" after the ¾" partition is installed above the cleats.

Glue strips of veneer to the raw front edges of the plywood sides, top and bottom. Stain and finish the enclosure shell to match your furniture or taste.

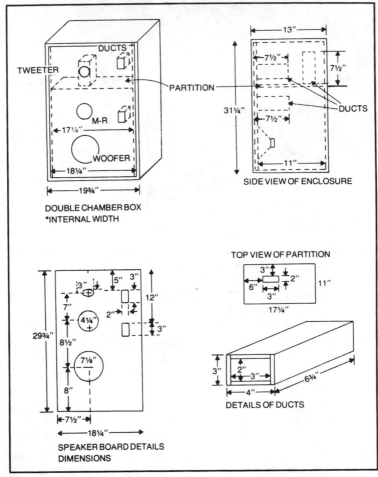

Fig. 5-16. Construction plans for double chamber reflex (Project 6).

Table 5-2. Parts List for Project 6.

¾" Plywood:

2	13" × 31¼"	Sides
2	13" × 19¾"	Top & bottom

Strip Veneer: ¾" wide:

about 9 feet	To cover edges of plywood

½" Particle Board:

2	11" × 18¼"	Liner, top & bottom
2	11" × 28¾"	Liner, sides
12	3" × 6¾"	Duct walls

¾" Particle Board:

2	18¼" × 29¾"	Speaker board & back panel
1	11" × 17¼"	Partition

¼" or ⅜" Plywood:

1	18¼" × 29¾"	Grille board

Speakers and Components:

1	8" woofer	Radio Shack Cat. No. 40-1006
1	Soft dome mid-range	Radio Shack Cat. No. 40-1281
1	Tweeter	Radio Shack Cat. No. 40-1374
1	2.2 μF capacitor	(supplied with tweeter)
1	2-way crossover network	Radio Shack Cat. No. 40-1296

Miscellaneous:

Fiberglass insulation	Radio Shack Cat. No. 42-1082
Grille cloth	

Make the ducts from ½" material so that they have interior tube areas of 2" × 3". Glue and nail the ducts to the prepared holes on the partition and speaker board. Drill a ¼" hole near the front of

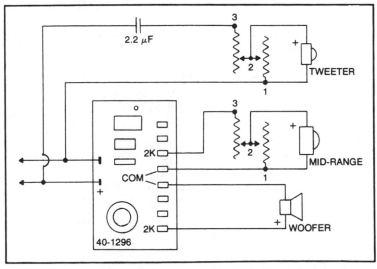

Fig. 5-17. Wiring diagram for Project 6, using Radio Shack 2-way crossover network, Model No. 40-1296.

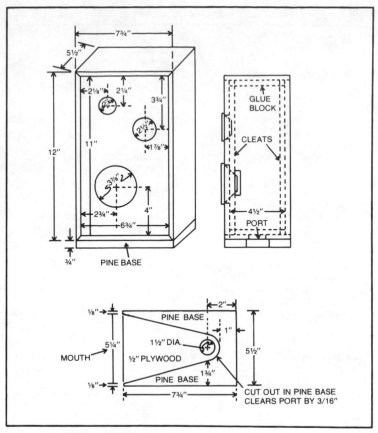

Fig. 5-18. Enclosure plans for Project 7.

the partition. Install the partition by gluing the matching surfaces and screwing or clamping it to the cleats. The partition adds greatly to the rigidity of the enclosure, front to back as well as side to side.

Next, install the crossover network. The crossover network is unusual because a 2-way network is used between the woofer and mid-range speaker. To use the 2-way crossover network to divide the signal to the woofer and mid-range speaker, remove the capacitor that came with the mid-range speaker. Connect the positive woofer terminal to the tap labeled "WOOFER 2K" and the negative woofer terminal to one of the taps labeled "COM." Connect pin #1 of the mid-range L-pad to one of the taps labeled "COM" and pin #3 of the L-pad to the crossover tap labeled "TWEETER 2K." Connect pin #1 of the L-pad to the negative terminal of the mid-range speaker and pin #2 to the positive terminal of that speaker. These connections will feed low frequency energy to the woofer and higher frequencies, from 2000 Hz upward, to the mid-range speaker.

The tweeter can be fed from the input to the crossover network through the 2.2 μF capacitor that comes with it, using an L-pad to control the tweeter's output level (Fig. 5-19)

The hook-up above is recommended, but you can save the cost of a crossover network by feeding the mid-range speaker and tweeter through the capacitors that are supplied with each speaker. Do get L-pads. As a refinement to this circuit you can add a homemade choke in series with the woofer. A 0.64 mH coil would be about right. Check Chapter 7 for details on making your own choke coil.

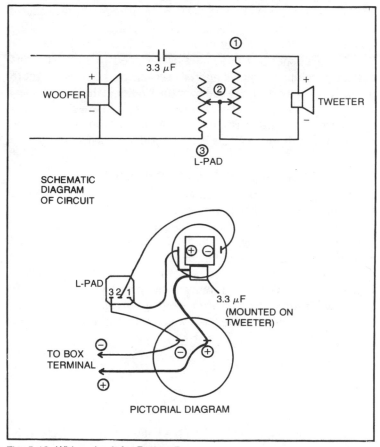

Fig. 5-19. Wiring circuit for Project 7.

Install the crossover components on an 8" × 8" crossover board and glue the board over a 4" × 4" cut-out in the back of the primary chamber. Feed the tweeter leads through the ¼" hole in the partition and fill the hole with caulking compound. Caulk all interior joints. Cover the interior of the primary chamber, the larger one, with fiberglass damping material. No damping material is needed in the smaller compartment. Cut the damping material away from the mouth of the port in the partition so that the port is clear of any obstruction.

Glue the speakers to the front of the speaker board with silicone rubber sealant. Allow the sealant to set over night before wiring the speakers, then use a generous amount of silicone rubber to glue the speaker board into the box. Make up a grille frame from ¼" or ⅜" plywood by making cut-outs at each speaker and port position. The cut-outs should be significantly larger than the size of the speakers or ports. Stretch a suitable grille cloth around the frame and install it by friction fit or with matching pieces of Velcro.

As mentioned earlier, there is no need to fine tune this enclosure because of the staggered resonances. But if you want to be able to install a damping pad over the woofer in the conventional way, by stapling it to the speaker board, you must alter the procedure slightly by installing cleats so you can screw in the speaker board. In our tests the damping pad was not necessary.

PROJECT 7: A COMPACT PORTED BOX SPEAKER SYSTEM

An optimum volume ported box will always be somewhat larger than the closed box for the same woofer (Fig. 4-8). But for small woofers, with a more limited bass range, the advantage of porting is more obvious. The 3 dB down frequency for Project 4 was about 110 Hz. When the same woofer is put into an optimum volume ported box, the cut-off point drops to about 66 Hz. Add an L-pad for more versatile room matching and you have a little system with surprising performance (Fig. 5-19).

The parameters of the small woofer, Radio Shack Model No. 40-1022, are:

$$f_s = 55 \text{ Hz}$$
$$Q = 0.35$$
$$V_{AS} = 0.23 \text{ cu. ft. or } 397 \text{ cu. in.}$$

Using Fig. 5-4, it appears that the volume factor for a Q of 0.35 is about 0.75. So to get V_B, multiply 0.75×397 cu. in., the speaker's V_{AS}, to get 298 cu. in. Rounding this off to 300 cu. in. and adding 10% overvolume, the required volume is about 330 cu. in. This is too small for the volume chart in Fig. 3-1, but a calculator gives 6¾" × 11" × 4½", for a final volume of 334 cu. in.

From Fig. 5-6 we find a tuning factor of 1.1. So to find f_B, multiply 1.1 × 55 Hz to get 60.5 Hz, the optimum tuned frequency of the box.

Construction

The only unusual feature of this project is the port. With such a small enclosure, it would be difficult to install a conventional duct large enough in diameter to avoid air noises but short enough to fit in the box. The port used here is based on what G. R. Koonce calls a "diffuser port." In this case the 1½" hole in the bottom of the box was cut to receive a section of 1⅜" I.D. plastic drainpipe, but in the final tuning the pipe was found to be unnecessary. The total port is formed by the enlarged opening that extends from the hole in the bottom near one side to the mouth at the other side of the enclosure (Figs. 5-18 and 5-20). The enclosure must rest on a flat surface when in use.

Make your stereo pair in mirror images of each other, revers-

Fig. 5-20. The pine base forms the expanded port for the enclosure of Project 7.

Table 5-3. Parts List for Project 7.

½″ Plywood		
2	5½″ × 12″	Sides
2	5½″ × 7¾″	Top & bottom
2	6¾″ × 11″	Speaker board & back
½″ × ½″ Wood Strips:		
4	½″ × ½″ × 6¾″	Cleats
4	½″ × ½″ × 10″	Cleats
4	½″ × ½″ × 3½″	Corner glue blocks
Pine:		
1	¾″ × 5½″ × 7¾″	Base
Grille Frame:		
As desired		
Speakers and Components:		
1	4″ Woofer	Radio Shack Cat. No. 40-1022
1	¾″ Hard dome tweeter	Radio Shack Cat. No. 40-1376
1	8 Ohm L-Pad	Radio Shack Cat. No. 40-977
1	Terminal plate	Radio Shack Cat. No. 274-625
Miscellaneous:		
Fiberglass insulation		Radio Shack Cat. No. 42-1082
Grille cloth		As desired

ing the position of the speakers on the second one. Don't expect to fill a gymnasium with sound from such small speakers, but, used within reason, they give exceptional performance for their size and cost.

6
Musical Instrument And PA Speakers

Audio fans occasionally try musical instrument speakers in their stereo systems just because they happen to be available. And, worse luck, musicians occasionally press high fidelity speakers into use. The outcome of such improvisations is usually disappointment for the audio fan and a blown speaker for the musician. The first rule of thumb involving high fidelity vs. musical instrument speakers is: they are not interchangeable. The same rule holds for PA speakers.

CHARACTERISTICS OF MUSICAL INSTRUMENT SPEAKERS

A musical instrument speaker must be durable and reliable. It must be able to take rough handling, as well as high power, without failing. Because of this requirement, musical instrument speakers have much stiffer suspensions than high fidelity speakers. While a stiff suspension restricts the cone movement and limits the extreme low frequency range, it doesn't significantly affect a musical instrument speaker's useful range. Such speakers aren't required to reproduce the sound of the lowest note of a huge pipe organ or similar instrument.

In addition to stiff suspensions, musical instrument speakers have reinforced cones, heat-proofed voice coils, and other features to prevent damage or burn-out. For still higher protection, these speakers are often used in pairs, or in groups of four, to handle more power.

PUBLIC ADDRESS SPEAKERS

The most important requirement for a public address speaker is that it can project the human voice with clarity, even in rooms with bad acoustic conditions, such as one with excessive reverberation. Extreme bass range is even less important for PA speakers than for musical instrument speakers. Deep bass response requires large speakers and the availability of more power. Another reason to limit bass response, low frequency dispersion is harder to control than that of the higher frequencies, and uncontrolled dispersion increases feedback.

The cue to good PA performance in a highly reverberant room is, again, controlled dispersion. One way to gain such control is to install column speakers, with 4 to 6 drivers in each column. A vertical column gives good horizontal dispersion but limited vertical dispersion. In a typical auditorium, with the speakers installed near the level of a stage or floor, vertical dispersion wastes sound energy by bouncing it off the floor and ceiling. Even worse, this reflected energy reaches the audience after a slight delay, reducing clarity. So PA columns are usually installed in a vertical position to limit vertical dispersion. In some narrow rooms, where the audience is compressed into a long rectangle with the speaker at one end, better audience coverage can be obtained by using the column in the ceiling, placed at right angles to the crowd pattern. The rule in PA work is: try everything.

AMPLIFIER-SPEAKER COMPATIBILITY

While almost all stereo amplifiers are solid state, many musical instrument amplifiers and PA amplifiers in use today are vacuum tube models. The rules for hooking up speakers to the two kinds of amplifiers are different. Here they are:

●**Tube Amplifier.** Speaker impedance should be equal or lower than that specified at amplifier output tap or jack. *Never operate amplifier without a load.*

●**Transistor Amplifier.** Speaker impedance should be equal or higher than that specified at amplifier output tap or jack. *Never operate amplifier with a low impedance or shorted load.*

Typical tube amplifiers will have several taps, or jacks, labeled 4, 8, and 16 ohms. For maximum efficiency the speaker system should be hooked to the output jack with the same impedance rating as the speaker. Some guitar players prefer to hook a 4-ohm speaker to an 8-ohm tap or an 8-ohm speaker to a 16-ohm tap because tube amplifiers seem to go into overload more smoothly when matched to a speaker of lower-than-normal impedance. When the mis-match occurs in the other direction, with a

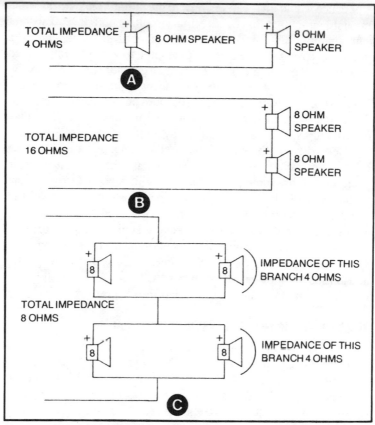

Fig. 6-1. How parallel and series wiring hook-ups affect impedance. (A) Parallel connection. (B) Series connection. (C) A parallel-series connection.

speaker of higher-than-normal impedance, overload is rougher, and if the load is infinite, such as happens with a broken speaker lead, the output tubes are usually blown.

Transistor amplifiers are rated for a minimum impedance load, and this rating must be observed. For example, if your transistor amplifier specifies a 4-ohm load, don't wire more than one 4-ohm speaker system to it unless they are wired in series (Fig. 6-1). If the amplifier has an 8-ohm tap, you would connect the series-wired speakers to that tap. If you have an amplifier with one tap, for an 8-ohm speaker, and you have two 8-ohm speakers that you want to use, you would wire them in series, making their total impedance 16 ohms. Such a mis-match would cause some reduction in power, but it would be safer for the amplifier because the higher impedance would limit current flow in the output transistors. Some transistor amplifiers have protection circuits, but it pays to protect your amplifier by correct matching rather than to depend on a protection circuit. Some users are tempted to wire low impedance speakers to their amplifiers because they get more power that way. A 2-ohm load in a 4-ohm output circuit will permit more current flow, so the amplifier tries to deliver more power than it is capable of handling without excessive heat build-up and eventual failure. When using a transistor amplifier, play it safe.

Regardless of the kind of amplifier you use, beware of pushing it into overload. One way that some musicians overload their amplifiers is by adding fuzz tones to the signal. Gadgets that produce fuzz do it by adding distortion, which can overload the amplifier or speaker system by an unnatural frequency balance. High frequency speakers are especially subject to such damage because a distorted signal will carry harmonics of the fundamental which have no relation to the normal frequency vs. power distribution of musical instruments. Operating an amplifier at levels above its rated power output has the same effect.

ENCLOSURES FOR MUSICAL INSTRUMENT SPEAKERS

The same laws of acoustics prevail for musical instrument speakers as for stereo speakers, but you wouldn't know it by inspecting many commercial musical instrument speakers. One big difference: a good percentage of commercial musical instrument speakers have no back on the enclosure. The reason seems to be the personal preference of some musicians for the kind of sound produced by an open box. One advantage of the speaker in the open box, a more pronounced bass response than is possible from the same speaker in a closed box of similar size. This gain in low frequency efficiency is obtained at the expense of power handling ability. If you put a speaker in an open box, you should limit the power to it to about half its rms rating. Open back speakers have one big disadvantage; when used with groups which include vocalists, or other performers with microphones, an open back speaker is more likely to cause feedback problems regardless of placement or orientation.

Plywood or particle board may be used in musical instrument or PA speaker enclosures. The typical thickness used in commercial enclosures is ¾", although ½" material is often used in small systems. The enclosure can be painted, but for a more professional appearance get some cloth-backed vinyl and glue it on the exterior.

Make sure the piece of vinyl is long enough to wrap around the enclosure from the bottom, up one side, across the top, down the other side, and under the bottom again. By using a continuous piece in this way there will be just one seam on the enclosure, under the bottom where it won't show. For protection and ease of handling, install metal corners and carrying handles. It's a good idea to mount large furniture glides under the box, or, for easy moving, heavy-duty casters. Such items can be found at most hardware stores.

While metal corners will protect the finish or veneer of your musical instrument enclosure, the speakers also need more protection than the typical stereo speaker. One way is to install hardware cloth, a heavy wire mesh, in front of each large cone. Such hardware cloth can be stapled to the rear of the speaker board, behind the speaker cut-out, before bolting the speaker in the box. Musical instrument speakers should be bolted to the speaker board. For easy bolting, mount T-nuts on the front of the speaker board at each mounting hole. Use the speaker as a pattern to locate the mounting holes, then drill the mounting holes using a bit a size larger than the bolt to be used; then drive in the T-nuts from the front. Use a split lock washer under the head of each mounting bolt. This kind of installation permits quick removal of the speaker without disturbing the grille cloth. When installing speakers with bolts, be careful not to overtighten the bolts. It is possible to warp the speaker, which will produce a rubbing voice coil.

Another way to protect musical instrument speakers while traveling is to short the voice coils. One easy way to do this is to keep a shorted plug handy for each enclosure. When you are ready to transport the speaker, disconnect it from the amplifier and insert the shorted plug into the jack on the back of the enclosure. When the voice coils are shorted, any cone movement must be done against the magnetic field, so such movement will be reduced.

If you travel with a group that is especially rough on speakers and other equipment, you can install plywood panels over the front of your speakers while they are not in use. Such a panel can be bolted to the speaker board via T-nuts mounted on the back of the speaker board. Don't forget to replace the bolts after removing the panel or there will be air leaks that might cause unmusical noise.

MONITOR SPEAKERS

After the first concert at a new small auditorium, some of the musicians complained that they couldn't hear the others playing. A musical instrument salesman told them the sound system didn't have enough power. I heard his recommendation of more power and knew it was wrong, so I checked with the musicians and found that the people at the rear of the stage couldn't hear because there were no monitors in the system. A small monitor solved the problem and proved that the power output was adequate.

Most musicians know the importance of a good monitor, but many speakers used as such are poorly suited for the purpose. Because a monitor speaker must face the performers, it often adds to the problem of feedback. The more vocalists or other performers with open microphones, the worse the problem. Careful placement of the monitor speaker can reduce feedback, but the monitor should be designed to cause a minimum of problems anyway.

There are two ways to design a monitor speaker for minimum feedback. These are accomplished by limiting the monitor's performance in two aspects: its direction of sound projection or its frequency response. If you are using a separate amplifier or channel for the monitor speaker, the frequency response can be controlled at the amplifier. By reducing bass range and level, feedback problems can be controlled. But direction of projection must be built into the speaker enclosure.

In planning a monitor system, you should also consider the

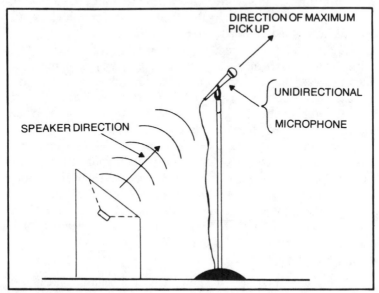

Fig. 6-2. One way to build a monitor speaker so that feedback problems are reduced.

kind of microphones you are using. Unidirectional mikes, such as Radio Shack Cat. Nos. 33-992 or 33-984, should be employed. The ideal monitor speaker will be designed so that its sound is aimed at the back of the microphone, where it is least susceptible to pick-up (Fig. 6-2).

Another solution to the monitor problem, especially useful where a monitor is needed for a single performer, is to install one or two small speakers in a mini-enclosure and mount the enclosure on a tripod or music stand close to the performer. An L-pad control on the monitor will permit the performer to adjust its volume so that he can hear the other performers through the monitor without blasting his ears or causing feedback.

PROJECT 8: A MUSICAL INSTRUMENT SPEAKER

The technique used in finishing a musical instrument speaker is different from that with a typical stereo speaker, but with a bit of patience you can produce a professional looking system. For this project, a 15″ musical instrument speaker and three piezoelectric super horns are required (Fig. 6-3). The super horns are installed on angled baffles to give good dispersion at high frequencies. Except for that feature, the cabinet work is typical of many other speaker systems and requires no special procedure.

Fig. 6-3. Musical instrument speaker with an expanded aluminum grille to protect the 15″ speaker in the lower compartment.

Construction

Cut out the parts according to the plans shown in Fig. 6-4 and Table 6-1. All angled cuts, such as those made on the outside tweeter boards and the cleats behind them, should be made at 15°.

In assembling the enclosure, glue every joint except when installing the back. The back should be installed with screws so it can be removed to service the large speaker if necessary.

Start assembly by gluing and nailing the bottom to the sides. Then glue and nail in the partition between the woofer and the tweeter compartments. Install cleats in the woofer compartment, setting them in ¾" at the rear and 1¼" at the front. Next, prepare the woofer board. Make the woofer cut-out in the board and then set

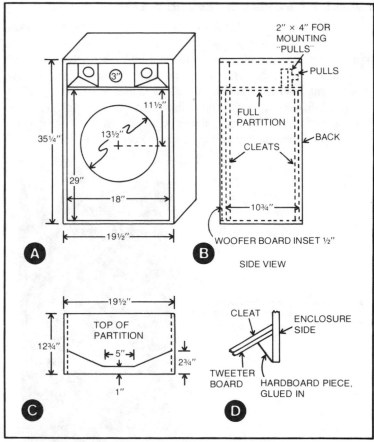

Fig. 6-4. Enclosure plans for Project 8. The lines on the partition (C) represent the boundary between the cleats and the tweeter boards.

Table 6-1. Parts List for Project 8.

¾" Plywood or Particle Board:

2	12¾"× 33¾"	Sides
2	12¾"× 19½"	Top & bottom
1	12¾"× 18"	Partition
2	18" × 29"	Woofer board & back

½" Plywood or Particle Board:

1	4" × 5"	Middle tweeter board
2	4"× 6⅞"	Outer tweeter boards

Pine:

4	1" × 2" × 18"	Cleats, woofer compartment
4	1"× 2"× 27½"	Cleats, woofer compartment
4	1"× 2"× 9¾"	Corner glue blocks, woofer comp.
1	½" × 1" × 5"	Middle cleat, tweeter board
2	½"× 1"× 6⅞"	Cleats, outer tweeter boards

⅛" Hardboard:

2	2½"× 4"	See plans

Expanded Aluminum:

1	18" × 29"	Grille for woofer

Speakers and Components:

1	15" Musical Instrument Speaker	Radio Shack Cat. No. 40-1315
3	Piezo super horns	Radio Shack Cat. No. 40-1381
1	20 Ohm resistor, 10 watt	
2	Terminal strips	Radio Shack Cat. No. 274-688
1	Terminal plate	Radio Shack No. 274-624

Miscellaneous:

	Fiberglass insulation	Radio Shack Cat. No. 42-1082
	Cloth backed vinyl, metal corners	
12	#6 × ½" Sheet metal screws	Tweeter mounting
8	1" × 3/16" Round head bolts	Woofer mounting
8	3/16" T-nuts	Woofer mounting

Optional Items for Carrying Handles:

1	2" × 4" × 18"	Mounting board for pulls
2	6-½" Brass pulls	Carrying handles
2	5" × 3/16" Bolts	To hold pull mounting board
8	2" × 3/16" Bolts	To hold pulls

the woofer over the cut-out and mark positions for the woofer bolts. Drill ¼" holes at each bolt position. Then drive 3/16" T-nuts into the holes at the front of the board. Install the board in the cabinet with glue and screws or nails.

Next, install cleats on the top of the partition, located to receive the tweeter boards. Note that the cleat for the center tweeter board is set back 1" from the front edge of the partition. A center line, drawn from front to back on the partition board, helps to center the cleat and the center tweeter board (Fig. 6-5).

Prepare and install the tweeter boards. Cut two pieces of ⅛" hardboard to 2½" × 4' dimensions. Install these pieces with glue at

Fig. 6-5. The tweeter board assembly for Project 8.

at each end of the tweeter boards, positioned so they extend from the inner front edge of each enclosure side, back to the tweeter boards (Figs. 6-4 and 6-5). Fill any gaps with caulking compound.

If you want to install carrying handles, cut an 18″ length of pine 2″ × 4″ and mount it near the rear of the tweeter compartment with two 5″ × 3/16″ bolts through the partition. This board should be positioned so that the pulls you mount on the board are accessible from the rear but will not extend beyond the rear edge of the partition or the top (Fig. 6-6). If you use particle board for the partition, put a ¼″ plywood liner under it to hold the nuts on the long bolts. Install the pulls on the 2″ × 4″ with eight 2″ × 3/16″ bolts.

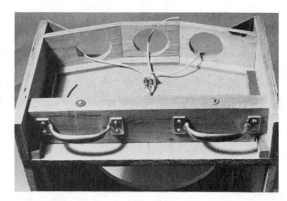

Fig. 6-6. Install tweeter wiring and optional carrying handles before putting top on cabinet.

Drill a ¼" hole through the partition and feed a length of lamp cord through it. Seal the hole with silicone rubber sealant. Install a small terminal strip behind the center tweeter board to receive the lamp cord from the woofer compartment. Make connections for the individual tweeter cords (Fig. 6-6). Remember to observe polarity in wiring.

Check your wiring to see that the proper connections are made and the joints are soldered with rosin core solder. Then install the top. Rasp or sand any rough exterior parts. Paint the front of the speaker boards flat black. Use an extra coat on the tweeter boards if necessary to obtain a smooth job. Cut a piece of black expanded aluminum to fit the front of the woofer compartment and glue it down with black GE Silicone II sealant. This metal grille protects the woofer from damage when moving the speaker.

Installing the Vinyl Covering

For a professional look, get a piece of cloth-backed vinyl material long enough to go around the box with no visible seams. The length from any point on the bottom of the box, up one side, across the top, and down the other side, back to the starting point, is 9 feet 1½ inches. To insure that the piece will reach around the box, add a few inches to the required length. This means that you must buy far more material than is required for a single enclosure, so if you don't mind a visible seam, you can get slightly more than half the required length and split the material lengthwise. To make the seam almost invisible, install the material with a slow setting glue, allowing the two pieces to overlap an inch or two. Then lay a straightedge, such as a large carpenter's square, across the over-lapped section and draw a razor blade along the straightedge. Lift the top layer of material to peel off the waste from the underneath piece. When this excess, as well as the top layer waste, is re-moved, you will have a closely matched seam that will barely show.

Because you will have more width than you need, you can cut the individual pieces in widths of about 18" in order to overlap the enclosure walls by at least 3" at the front and by 1½" at the rear. After gluing these strips around the edges of the boards, trim the excess material away with a razor blade.

As mentioned earlier, it is easier to make seams with a slow setting glue, such as ordinary white wood glue. The preferred glue for this kind of work is contact cement, but it requires careful alignment of sections before contact with the box is made. This procedure is difficult when using such large pieces of material. For the enclosure shown in the pictures, wood glue was used on the broad sides, and then, after that glue had set, the overlapping edges of material were glued down with contact cement. This procedure permits careful alignment of the large sections, plus the straightedge and razor cutting method for the seam. The material can be folded at the corners to leave a 45° fold down each corner. Notching the material there will reduce the bulge made by simple folding, and the metal corners will cover any minor mistakes.

The only other difficult matching is in cutting a piece of vinyl to cover the front edge of the shelf which separates the tweeter boards from the large speaker board. Cut a piece of vinyl about 4 inches wide and long enough to overlap the ends of the pieces in place. Use the straightedge and razor technique to match the seam edges, but work quickly before the contact cement has set. The relatively wide strip is suggested so you can trim off the excess at any point you desire. For example, you can let the vinyl extend into the base of the tweeter compartment, covering the top of the shelf.

Install the speakers, using 3/16″ × 1″ bolts for the large speaker and #6 × ½″ sheet metal screws for the tweeters. Note that the tweeters are installed from the front of the boards, and the large speaker is from the rear. In wiring the tweeters, note that the

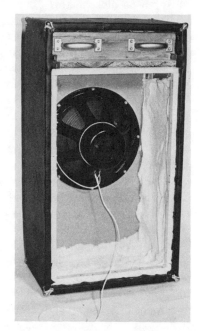

Fig. 6-7. Cleats permit access to woofer compartment through screwed-on back.

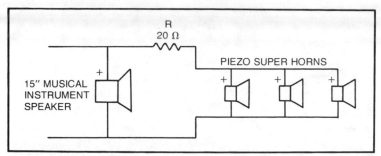

Fig. 6-8. Wiring circuit for Project 8.

positive terminal is on the right solder lug when you look at the tweeter from the back.

No crossover network is needed, but a 20- to 30-ohm resistor should be wired in series with the tweeters (Fig. 6-8). The tweeters are wired in parallel, which causes no problems with most amplifiers, but some amplifiers are "capacitive sensitive" at about 40 kHz. The precautionary resistor prevents such an amplifier from going into oscillation at high frequencies.

To mount the series resistor, install a terminal strip in the woofer compartment. Connect the 20-ohm, 10-watt resistor to the solder lugs. Then break one of the lamp cord conductors from the tweeter compartment and connect the broken ends to each of the lugs. Solder with rosin core solder. Connect the ends of the lamp cord to the woofer. Then connect another piece of lamp cord to the woofer lugs, long enough to reach the terminal strip in the enclosure back (Fig. 6-7).

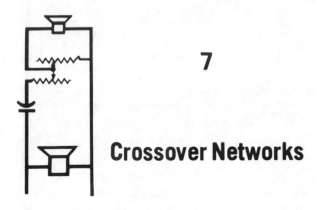

7

Crossover Networks

A single cone extended range speaker has advantages, such as low cost and easy wiring. It also has some shortcomings, limited dispersion at high frequencies and somewhat greater distortion than a 2-way or 3-way system. When a single cone is expected to reproduce the entire frequency range, there is a good chance that the large cone movements necessary to produce the bass will modulate the treble. An obvious way to reduce this kind of distortion is to use one speaker for bass and another for treble, a woofer-tweeter combination. Or, for larger woofers, a mid-range driver is added for a 3-way system. This method of covering the frequency range permits us to use speakers that are specially designed to give optimum performance in the frequency bands assigned to them.

If you want to build a 2-way or a 3-way system, you must make sure that the frequencies fed to each driver are right for it. For example, it is essential that the highs go to the tweeter and, more important, that lows are blocked from it. If a small tweeter receives very much low frequency power, it will produce severe distortion, then fail. A major requirement for multi-way speakers is a properly designed crossover network.

KINDS OF CROSSOVER NETWORKS

Commercial speaker systems have a wide variety of crossover networks, from those with a single crossover component to some with extremely complicated circuits. The simpler networks sometimes perform better than the more complex ones because of more phase shift in the higher order networks. The simplest kind of crossover network is a single capacitor, called a *high-pass filter.*

High-Pass Filters

A *capacitor* offers high reactance to low frequency energy, so a capacitor in series with a tweeter serves as a high-pass filter(Fig. 7-1). A capacitor passes the highs but restricts the lows. Capacitance is measured in farads, but the values used in crossover networks are measured in microfarads (μF) or millionths of a farad. At the crossover frequency the capacitor's reactance is equal to the impedance of the tweeter. For example, if you have an 8-ohm tweeter that you want to use with a 4000 Hz crossover frequency, you would choose a capacitor that gives a reactance of 8 ohms at 4000 Hz. To find the correct value of capacitance, go to Fig. 7-2 and follow the vertical line up from 4000 Hz to the horizontal line for 8-ohm speakers. Then move diagonally to the upper left on the dashed line, and you will find that a 5 μF capacitor will provide the proper impedance at the desired crossover frequency. A 4.7μF capacitor, available at Radio Shack stores, will work just as well as one rated precisely at 5 μF.

When you buy a capacitor for this purpose, make sure it is a non-polarized electrolytic or a mylar type. The non-polarized electrolytic kind is available in larger values and is considerably less expensive than large mylar capacitors. Don't use polarized electrolytics for crossover duty.

If you can't find the value you need in a single capacitor, you can use more than one to obtain the desired value. Remember this rule: when capacitors are wired in parallel, their values are added. For example, if you need a capacitance of approximately 15 μF and you have 10 μF and 4.7μF capacitors, you can wire one capacitor of each of these values in parallel.

Low-Pass Filters

A *coil* has the opposite frequency characteristic to that of a capacitor. Thus a coil in series with a woofer chokes off the high frequencies because its reactance to an alternating current increases with the frequency of the current. The more turns in a coil, the greater its reactance at any specified high frequency. One of the most popular kinds of crossover networks consists of a coil in series with the woofer and a capacitor in series with the tweeter (Fig. 7-3).

109

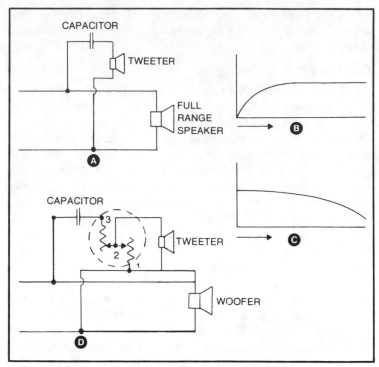

Fig. 7-1. A single high-pass filter is the simplest kind of crossover network. (A) A capacitor acts as a high-pass filter, and a tweeter is added to a full range speaker to extend high frequency range. (B) Tweeter response adds high frequency range to that of full range. (C) A full range speaker has a dropping high frequency response. (D) Adding an L-pad in the tweeter circuit helps to balance the output of the tweeter to that of the woofer.

That property of a coil which discriminates against highs is called *inductance,* and coils are often called *inductors.* Inductance is measured in henries, but the values that are useful in crossover networks are more conveniently measured in millihenries (mH) or thousandths of a henry. To find the correct value of inductance for a crossover network, use the chart in Fig. 7-2 as you did for capacitance, but for inductance you must follow the dashed line to the upper right. For example, if you want to make a low-pass filter for an 8-ohm speaker with the crossover point at 2000 Hz, you would find the 2K line on the chart, follow it up to the horizontal line for 8 ohms, then upward to the right to 0.6 mH.

There are two kinds of choke coils that can be used, *air core* and *iron core.* Most commercial crossover networks use iron core chokes because they conserve wire and have lower DC resistance.

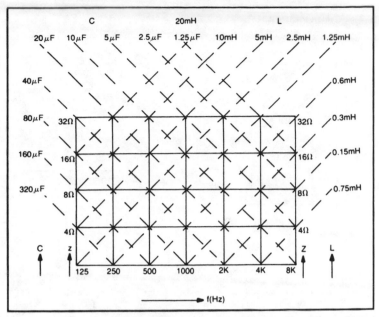

Fig. 7-2. Crossover network design chart. Approximate values of inductance and capacitance for speakers with ratings from 4 to 32 ohms.

An air core choke needs more turns of wire for the same inductance, but if you can find some magnet wire, usually available from motor repair shops, you can make your own air core chokes. Instructions will be given later in the chapter.

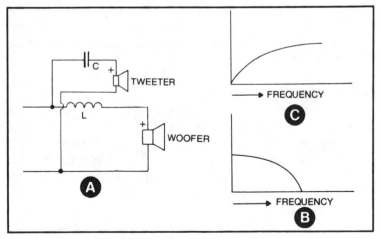

Fig. 7-3. A coil in series with the woofer chokes off the high frequencies.

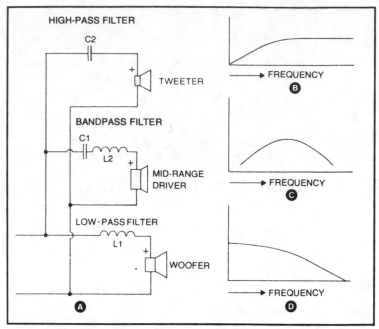

Fig. 7-4. (A) A parallel 3-way crossover network with 6 dB per octave roll-off. Many commercial crossovers available for speaker builders use this kind of circuit. (B) Output of tweeter. (C) Output of mid-range driver. (D) Woofer output.

Bandpass Filters

When a capacitor is placed in series with an inductor, the combination forms a *bandpass filter* (Fig. 7-4). The values of the two components must be calculated so that the capacitor blocks the lows below the frequency at the lower end of the desired band width and the coil blocks the highs beyond the upper end of the desired bandwidth.

First Order Networks

Crossover networks are sometimes described by terms such as *first order, second order,* and so on. A first order network is one that causes the energy to a driver to be rolled off at the rate of 6 dB per octave beyond the cut-off point. This kind of network has a single component in each branch of a 2-way network (Fig. 7-3), but in a 3-way network it has 2 components in the bandpass filter for the mid-range driver.

First order networks are preferred by many engineers because they are uncomplicated, economical, and have desirable phase characteristics. Their only disadvantage is that they permit drivers to produce sound far beyond the crossover point. This is not a great shortcoming unless the drivers have peaks in their response curves that should be eliminated. The one exception is for a small tweeter that is operated down to the bottom of the safe frequency range of the tweeter. A first order network, with just a single capacitor in series with the tweeter, may permit enough low frequency energy to pass through the tweeter to damage it. The solution is to move the crossover point higher or, if that isn't practical, to use a *second order network* for the tweeter alone.

Second Order Networks

Second order networks have two components in each branch of the circuit and produce a 12 dB per octave roll-off beyond the crossover frequency. For example a tweeter would have a capacitor in series with it and a coil in parallel (Fig. 7-5). To design a second order network for your tweeter, look up the value of capacitance for the crossover point you want to use, then multiply it by 0.7. Then find the value of inductance for the same frequency and multiply it by 1.414. For a second order filter for an 8-ohm tweeter with the crossover point at 4000 Hz, the correct values to

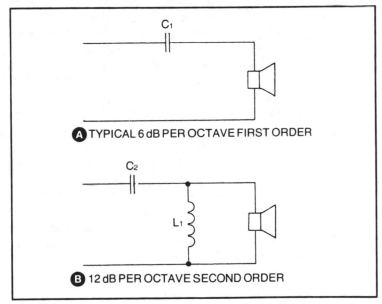

C_1

Ⓐ TYPICAL 6 dB PER OCTAVE FIRST ORDER

C_2

L_1

Ⓑ 12 dB PER OCTAVE SECOND ORDER

Fig. 7-5. Changing the value of C and adding a choke across a tweeter provides a sharper cut-off at the crossover point.

Table 7-1. Values of C_2 And L_1 for Circuit Shown in Fig. 7-5.

Crossover Frequency	Original Capacitor Value (μF)	Capacitor Value To: (μF)	add L with a value of: (mH)
5000	4	2.8	0.36
6000	3.3	2.3	0.3
7000	2.8	2	0.26
8000	2.5	1.8	0.22
9000	2.2	1.6	0.2
10000	2	1.4	0.18

use would be 3.5 μF and 0.42 mH. These would provide the same crossover point, 4000 Hz, as a single 5 μF capacitor, but the roll-off is steeper. Table 7-1 gives some common values for second order tweeter circuits.

Many crossover networks are hybrids, containing both first order and second order sections. Most second order 3-way networks have a first order bandpass section for the mid-range.

HOW TO CHOOSE CROSSOVER FREQUENCIES TO MATCH DRIVERS

The easiest way to choose the right crossover frequency for a driver is to follow the manufacturer's recommendations. If a woofer is listed as having an effective frequency range to 2500 Hz and you want to use it with a mid-range driver that has a range from 1500 Hz to 10,000 Hz, then you can choose any frequency between 1500 Hz and 2500 Hz for the woofer to mid-range crossover frequency. You can always stretch a woofer's range with no damage to the speaker system, although there may be a slight dip in the upper bass or lower mid-range. But be careful about using mid-range drivers or tweeters below the frequency range suggested by the manufacturer. It can be done safely only if the system is to be used for low power applications. With some drivers the sound quality may suffer if you permit them to operate through their frequency of resonance. As a rule of thumb, mid-range speakers and tweeters should be used with crossover networks that cut off energy to the driver an octave or more above their fundamental resonance. For example a mid-range speaker with a resonance at 500 Hz should be used with a crossover frequency of 1000 Hz or higher.

Theoretically a woofer should be used to cover only that frequency range below where the wavelength of the sound is equal to the effective cone diameter of the woofer. This would be a rigid rule only if speaker cones were rigid pistons. They obviously are not, but here are the theoretical upper limits: 8" woofer, 1500 to 2000 Hz; 10", 1500 Hz; 12", 1200 Hz; and 15", 1000 Hz.

COMMERCIAL CROSSOVER NETWORKS

A minimal crossover network, such as a high-pass filter, is useful for many speaker systems, but a full crossover network is more desirable. A large woofer, such as a 12" model, usually gives better performance if the crossover frequency is lower than that provided by a typical general purpose network. This problem, and several others, can be solved with a crossover network that offers a choice of crossover frequencies.

For example, Radio Shack's 3-way crossover network, Cat. No. 40-1299, has tapped chokes and multiple capacitors so that the user can choose the correct tap or the right capacitor to match the requirements of his speakers. If this sounds complicated, it isn't at all. You simply wire the output of your receiver to the two lugs marked "IN," one terminal of your woofer to the common, or negative, lug on the crossover network, and the positive terminal of the woofer to either of the lugs marked 800 Hz or 1600 Hz, as desired. If you choose 800 Hz for the woofer, a logical choice for a 12" speaker, then you would wire the mid-range speaker to the 800 Hz to 5000 Hz band and the tweeter to the 5000 Hz crossover. For a smaller woofer or a mid-range speaker with a more limited low end response, you would choose the 1600 Hz tap for the woofer, the 1600 Hz to 7000 Hz band for the mid-range speaker, and the 7000 Hz crossover point for the tweeter. There are 3 wiring lugs for each driver, a common lug and a choice of the other 2 lugs to select the desired crossover frequency. This makes the crossover network both versatile and easy to use. It is a suitable choice for any of the 3-way projects in this book. Choose the 1600 Hz and 7000 Hz crossover frequencies. But make sure you remove any capacitors from mid-range drivers or tweeters before connecting them to this crossover network. Failure to do that would put two capacitors in series, decreasing the value of capacitance and increasing the crossover frequency.

Radio Shack's 2-way crossover network, Cat. No. 40-1296, also has a tapped choke and multiple capacitors. For this unit the crossover frequencies are 2000 Hz, 2500 Hz, or 4000 Hz. With a choice of 3 crossover frequencies there are 4 connecting lugs for each driver, a common lug and a choice of 3 lugs for the other speaker terminal (Fig. 7-6).

Fig. 7-6. Two variable frequency crossover networks. Tapped chokes and multiple capacitors permit choice of crossover frequencies for 3-way network, above, or 2-way network, below.

EXPERIMENTS WITH MULTI-TAPPED CROSSOVER NETWORKS

The crossover networks described above are designed to be used in a straightforward manner, that is by wiring the woofer and the tweeter or the woofer and the mid-range speaker to share the same crossover frequency. But they don't have to be used in a rigid manner. Most speakers, regardless of quality or price, convert electrical energy to sound more efficiently in the mid-range than at other frequencies. The ear, also, is more efficient at converting sound energy to nerve impulses to the brain at the middle frequencies, particularly when the sound is reproduced at low volume levels. Because of these peculiarities of speakers and the ear, it is sometimes better to wire a woofer to the 800 Hz tap on the crossover network and the mid-range speaker to the 1600 Hz tap. Or, in the 2-way network, choose the 2000 Hz tap for the woofer and the 2500 Hz tap for the tweeter. If this seems like heresy, remember that crossover networks don't act like a stone wall. Instead of chopping off the response beyond the crossover frequencies, they produce gradual attenuation. Sometimes you can change a good speaker system to a great speaker system by just this kind of experimentation. It's one of the advantages of building your own speakers.

Here is another unusual hook-up. Suppose you want to use Radio Shack's new mid-range tweeter in a 2-way system with a 10" or 12" woofer. This mid-range tweeter, Cat. No. 40-1299, can be used down to 700 Hz, so an 800 Hz crossover would be a good choice. You can convert the 3-way crossover network into an 800

116

Hz 2-way network by running a wire from the positive "IN" terminal to the closest lead on the 24 μF capacitor. This by-passes the small coil which limits high frequency response when used in series with a mid-range speaker.

A HOMEMADE CROSSOVER NETWORK

Many tweeters and mid-range speakers are marketed with a suitable crossover capacitor included with the driver. By adding a single coil you can make a simple 3-way crossover network that consists of a low-pass filter for the woofer and high-pass filters for the other drivers.

The typical value of the capacitor supplied with mid-range drivers is 15 μF, and for tweeters, 4 μF. These values give crossover frequencies of about 1300 Hz for the mid-range speaker and 5000 Hz for the tweeter. With most woofers you can use the high-pass filters alone and get satisfactory performance, but a 1 mH choke in series with the woofer conserves power and avoids undue overlapping of woofer and mid-range output in the mid-range. You can make such a coil with about 69 feet of #20 gauge magnet wire and a suitable form (Fig. 7-7 and Table 7-2). This crossover network was used in several of the projects in this book.

HOW TO MAKE A CHOKE COIL

If you want to make your own coils, get some magnet wire from a motor repair shop, then make some forms and wind the coils by hand or with an electric drill. You must first find the value of inductance that you need from Fig. 7-2. Then go to Table 7-2 for coil dimensions and proper wire gauge. Or, if the inductance you need is not listed in Table 7-2, you can use the graphs in Appendix C.

When you have the correct dimensions, make a coil form. For a permanent form you can use a piece of wood dowel of the right diameter for the core and pieces of hardboard for the sides. Glue the sides to the core. Don't use iron screws, nails, or any other iron parts in the form because they would increase the inductance unpredictably by changing the coil from an air core, or no iron, to an iron core choke.

A temporary form is one that you can use to wind the coil, then take apart to remove the sides and use them for another form. The difference is that you make up fewer sides, but you must be careful to tape the coil so it will hold its shape. The core should be hollow in a temporary form so you can leave the wire on the core and tape

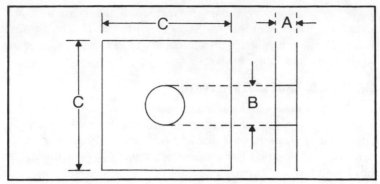

Fig. 7-7. Coil form dimensions. To find correct dimensions, refer to corresponding letters and figures in Table 7-2.

the coil radially, running the tape through the core and out around the outside of the coil. Pieces of plastic pipe or plastic pill bottles make good cores for temporary forms.

To prepare the form, cut out the sides and drill a small exit hole just outside the position to be occupied by the core. Then glue the form, or bolt it together. If you are planning to wind the coil with an electric drill, you will want to use a ¼" bolt through the core regardless of whether the form is a permanent one or temporary. The bolt can be used as a clamp to hold the sides of a

Table 7-2. Wire And Form Data for Some Frequently Used Coils for Crossover Networks. For Coil Form Symbols, Refer to Fig. 7-7.

Frequency (8 ohm) speaker)	L (mH)	Wire Length (ft.)	No. Turns	Wire Gauge	Form Dimensions
10000	0.127	17.5	89		A = ¼"
7000	0.18	20.8	106	#24	B = ½"
5000	0.25	24.5	125		C = 1"
4000	0.32	27.7	141		
3500	0.36	36	122		A = ⅜"
3000	0.42	39	132	#22	B = ¾"
2500	0.51	43	145		C = 1½"
2000	0.64	56	141		A = ½"
1500	0.85	64	162	#20	B = 1"
1000	1.27	78	200		C = 2"
900	1.41	101	171		A = ¾"
700	1.82	115	194	#18	B = 1½"
500	2.55	136	230		C = 3" to 4"
300	4.24	175	296		

permanent form until the glue sets. The bolt should be long enough to reach through the form far enough to hold washers on each side, two nuts, and an extra half inch or so to fit into the drill chuck. Two nuts are needed so one can be jammed against the other to prevent the form nut from unscrewing as you wind the coil. Don't forget that you must remove all iron from the coil before using it.

Coil specifications usually list the number of turns to make a coil with the correct inductance. Tests on many coils suggest that it is easier and more reliable to measure the length of wire. Make sure the coil dimensions and the gauge of wire is right.

To wind the coil, thread a few inches of wire through the center exit hole, from inside to outside. Note that you should add enough to the specified wire length to include this extra wire plus one at the outside of the coil. Begin winding, making sure that the first loop is placed tightly against the side of the form with the exit hole. Try to make each turn lie flat against the previous one. If you stop winding for any reason, wrap the part of the coil already wound with plastic tape to hold the wire in place until you start winding again.

When you have completed the winding, tape the coil around the exposed surface of copper wire to hold the turns tightly in place. If you are using a temporary form, remove the nuts and pull off the coil sides. Tape the coil by winding the tape through the hole in the "doughnut" and over the outer circumference.

Don't be intimidated by the coil winding process. If you are even reasonably careful in following specifications, the coil winding will be successful. Even a scramble wound air core coil will perform well unless the scrambling is extremely sloppy.

L-PADS

Tweeters and mid-range drivers are usually more efficient than large woofers, so there must be some method of controlling the sound level of these smaller speakers. Ordinary fixed resistors can be used to balance the sound of the various drivers in a speaker system, but the value of each resistor would have to be carefully chosen. Another problem with fixed resistors; you can't change the level of a driver to compensate for the acoustic conditions in your room. And, unless consideration is given to keeping the proper impedance for the crossover network, fixed resistors can cause a shift in the crossover frequencies.

These objections can be overcome by using an L-pad. An L-pad adjusts the sound level of the driver it controls much like a volume control, but it maintains a constant impedance for the crossover network and amplifier. It does this by varying a series resistance inversely with a parallel resistance (Fig. 7-1). Since almost all high fidelity speakers have a rated impedance of 8 ohms, most L-pads present a load of 8 ohms.

Choose your L-pad to match the power and crossover frequencies of your system. The power level at high frequencies is always much reduced from those of the mid-range, so you can use an L-pad designed for lower power handling in most tweeter circuits or for systems to be operated at low sound levels.

While some systems work perfectly well with no level controls on any of the drivers, most systems can be improved by adding L-pads to mid-range and high frequency speakers. In fact, one of the real advantages of a 3-way system over a 2-way system is that you have better control over the sound level at various frequency bands.

Use great care in adjusting the controls on your mid-range or tweeter drivers. The best way is to begin by turning the levels of each all the way down. Then bring up the mid-range level until it balances and blends with that of the woofer. Do the same with the tweeter control. Try to make the system sound as much as possible like a single speaker.

CONTOUR NETWORKS

In addition to the standard crossover networks audio engineers sometimes make use of special circuits to gain more control over a speaker system's response. A network that shapes the speaker's response curve is called a *contour network*. It is typical of many full range speakers to be most efficient in the band where the human ear is most sensitive, in the mid-range. Many listeners prefer this band to be slightly depressed. With a 3-way system and an L-pad, you can often achieve the desired effect by turning down the volume on the mid-range, but what do you do for a single-cone speaker? The answer is the contour network.

When a choke and a capacitor are wired in parallel, they form a circuit that is resonant at the frequency where the inductive reactance equals the capacitive reactance. At this frequency the current in the choke is equal to that in the capacitor but out of phase with it. If the circuit were composed of a pure capacitance and a pure inductance in parallel, no current would circulate in other parts of the circuit at the frequency of resonance, and the circuit's impedance at that frequency would be infinite. The capacitor and choke would form an isolated tank circuit in which a series current would flow back and forth. Note that this current appears only in the tank circuit at resonance, not in the line outside the tank.

If you wanted to eliminate sound at a certain frequency, you could use such a tank circuit, but with real components it wouldn't be perfect. A choke, for example, always includes some resistance in addition to its inductance. For a practical circuit, a third component should be added: a resistor. When a parallel resistance is wired into the tank circuit (Fig. 7-8) the resistor carries some current through regardless of frequency. If the value of resistance is several times the reactance of the other components at the frequency of resonance, the resistor will make little difference to the current flow in the circuit. But if it is about the same or lower in resistance, it will broaden the frequency response of the filter.

To design a contour network, imagine that you are designing a 2-way crossover network for speakers that have twice the impedance of your speaker. The most likely frequencies are 1000 to 2000 Hz. For example, suppose you want to make a filter for an 8-ohm speaker, centered at 2000 Hz. Using Fig. 7-2, we move up the 2K line to the horizontal line for 16-ohm speakers. From that point we find that the capacitor should be 5 μF and the choke, 1.25 mH. We would wire these components in parallel, then parallel them with a 16-ohm resistor. To get more depression of the frequencies around 2000 Hz, we would try a larger resistor; for a broader, shallower-acting filter, a smaller resistor.

One precaution about using such a filter; if you try the filter by shorting around it, to remove it from the circuit, the increase in loudness without the filter may sway your judgment against it. Most people will prefer the louder of any two speakers. If you must test the filter, try to adjust the volume control simultaneously to make up for the loss in sound level.

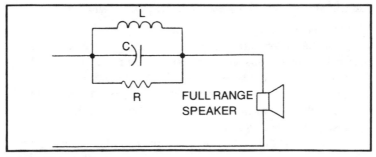

Fig. 7-8. A contour network.

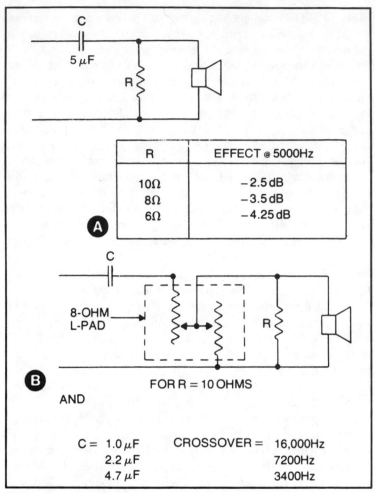

R	EFFECT @ 5000Hz
10Ω	−2.5 dB
8Ω	−3.5 dB
6Ω	−4.25 dB

FOR R = 10 OHMS

AND

C =	1.0 μF	CROSSOVER =	16,000Hz
	2.2 μF		7200Hz
	4.7 μF		3400Hz

Fig. 7-9. Low end roll-off circuits for piezoelectric tweeters.

CROSSOVER NETWORKS AND PIEZOELECTRIC TWEETERS

One of the great advantages of piezoelectric tweeters is that they can be wired into a speaker system without a crossover network. The impedance of these tweeters is so high at low frequencies that the tweeter is effectively out of the circuit. Because of this feature piezoelectric tweeters can be wired into any speaker system by hooking them directly to the amplifier, bypassing any crossover network components.

If you want to limit the lower response limit of these tweeters, you can add a capacitor and resistor for a low end roll-off at 3 dB per octave (Fig. 7-9). To put a variable control on the tweeter, you can use the other circuit in Fig. 7-9. In each of these circuits the resistor and capacitor form a network that controls the roll-off point. This permits you to use the piezoelectric tweeter purely as a super tweeter, if desired.

PROJECT 9: THREE-WAY SYSTEM WITH 12″ WOOFER

This system uses the same enclosure volume, as well as the same mid-range driver and leaf tweeter, as Project 5 (Fig. 7-10 and Table 7-3). The 12″ woofer here costs a bit more, but your choice between the two projects is more likely to be made on the basis of ported versus closed box than on cost. If you like your music super loud, choose this one. The closed box and the larger woofer give a higher safety margin.

Construction

Make the woofer cut-out as shown in Fig. 7-11. That and the absence of a port are the only differences in this enclosure from the one described in Project 5. You can build the box from plywood with beveled or dadoed joints, as shown, or choose cheaper particle board and use butt joints. Install cleats to hold the speaker board and back panel, and put glue blocks inside each corner joint to improve rigidity. Caulk all joints. Add a brace to the back, installed as described for Project 5.

Another view of Project 9 is shown in Fig. 8-1. There you can

Fig. 7-10. The 12″ woofer gives a full bass response. System resonance occurs at 45 Hz.

Table 7-3. Parts List for Project 9.

Speakers and Components:

1	12" Polypropylene Woofer	Radio Shack Cat. No. 40-1023
1	4" Mid-range driver	Radio Shack Cat. No. 40-1282
1	Leaf tweeter	Radio Shack Cat. No. 40-1375
1	3-way Crossover network	Radio Shack Cat. No. 40-1299
2	8-ohm L-pads	Radio Shack Cat. No. 40-977
1	Terminal plate	Radio Shack Cat. No. 274-625

see the treatment given the plywood sides, a finish that some people like. If you want that much grain, burn the enclosure with a propane torch before varnishing or oiling it. A slight scorching raises the contrast of the wood grain and can make minor flaws less obvious. If you try this, first practice with a piece of waste material. Make sure your work place is free of combustibles. After burning the walls you can follow with your usual finishing techniques.

As in Project 5, the crossover network goes behind the speaker board near the mid-range driver and tweeter. Remember to install a 1" layer of fiberglass on the walls and back.

Conclusion

If you compare the three closed box projects of greatly different sizes, Projects 3, 4, and 9, you will find that the smallest system has the lowest Q and the largest system has the highest Q. There is a valid reason for this divergence. Small systems have higher frequencies of resonance, so the Q should be low to prevent

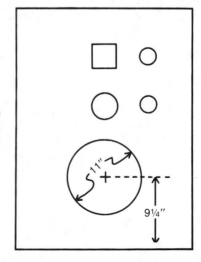

Fig. 7-1. Speaker board for Project 9. Except for location and diameter of woofer cut-out, all enclosure dimensions are identical to those of Project 5.

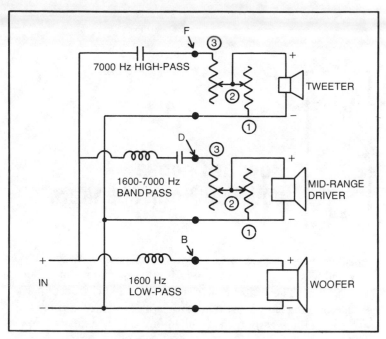

Fig. 7-12. Wiring diagram for Projects 5 and 9. Letters B, D, and F refer to solder lugs on 3-way crossover network, Radio Shack Model No. 40-1299. Numbers 1, 2, and 3 refer to connection points on L-pads.

peaking within the range of the human voice. For Project 9 the final frequency of resonance is 45 Hz. At such a low frequency the Q of 1.7 is hardly noticeable.

Rock music fans sometimes favor even a higher Q at the risk of unnatural bass. Reducing the box volume to about 2 cubic feet will raise the Q of this woofer to 2 or higher. But—double jeopardy—it will also put the system resonance above 50 Hz. Unless you are sure about what kind of bass freak you are, stick with the design shown here.

A small speaker stand that elevates the cabinet a few inches and tilts it back several degrees will provide a smoother bass response and better treble distribution.

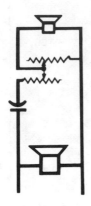

8

How to Choose and Use Your Speakers

A speaker that has good tonal balance in one room may sound too bassy and hard in a second room and either shrill or muffled in a third location. These apparent changes in sound character occur because of different acoustic conditions in the various rooms. A room with a thick carpet, overstuffed furniture and heavy drapes can absorb much more mid and high frequency energy than one with a tile or hardwood floor, large uncovered glass areas and unpadded furniture.

The ideal listening room will have some absorptive surfaces, but if too much area consists of soft surfaces, the ratio of direct to reflected sound will be increased to an unnatural degree. This can make the highs harder to take even though the high frequency tones may be reduced in output level. A hard room increases the apparent mid-range response, bringing voices and some instruments forward so they dominate the sound. Some of these effects can be countered by judicious use of the mid-range and tweeter controls. If you are using a full-range single-cone speaker, you can often improve its sound by use of a contour network, such as those described in Chapter 7. Regardless of the kind of speaker, you can use a sound level meter and a graphic equalizer, as described in Chapter 9, to correct for bad room acoustics.

There is no way to predict with certainty how any set of speakers will perform in a specific location, but you can put the odds in your favor by systematically considering how you will use your speakers before you choose them. And, after making the choice, there are ways to insure that you get the best possible performance from those speakers.

CHOOSING A SPEAKER

To pick the right speaker you must consider the other components you will use, the kind of music you like, the number of speakers you intend to employ, and the kind of room you will put them in, but not necessarily in that order. To narrow the choice, first decide the size of speaker system that will be appropriate. Then consider the other characteristics.

SIZE OF SPEAKER VS. SIZE OF ROOM

The rule on choice of speaker size is to keep things in proper scale. You obviously wouldn't want to try to fill a gymnasium with sound from a pair of mini-speakers. Nor would you try to put a couple of large floor models in a van. But many audio fans make the mistake of choosing a speaker size that is inappropriate for the job it must do. How can you judge the proper size?

Fortunately there is a rule of thumb that works well. If the proportions of speaker size to room size look right, they are probably right acoustically. To test your initial ideas, make up a mock speaker by cutting and taping together some pieces of shipping cartons. If the carton appears to be lost among the rest of the furniture, it is probably below optimum size for the room. If it overpowers everything around it, it's probably larger than necessary. But if you can arrange the carton so it fits nicely into your room and with the other furniture, chances are that you've found the right size of speaker for satisfactory bass reproduction without wasting space.

If this method of choosing optimum speaker size seems mystical, it really isn't. It is based on the fact that large rooms require much greater low frequency power output than small rooms. Large speaker systems have larger woofers that can pump out a higher level of bass response without distortion. No quick rule of thumb is 100% infallible, but this one will work in most situations. It can be modified by the kind of music you favor.

MUSICAL TASTES

Almost any size of speaker can reproduce the tonal range and volume of a string quartet or supply easy listening background music, but a rock group or a symphony orchestra played back at high sound levels requires more output power than a small woofer can produce. If you like your music really loud, you probably need a speaker with a woofer that is 10″ or 12″ in diameter. The tonal range of rock music can be easily handled by a smaller woofer, but not at high sound intensity levels.

USING YOUR SPEAKERS

Even if you choose speakers that are adequate for the job, you can be disappointed in their sound. There are several aspects of the problem of optimum use, but first make sure that your speakers are properly installed. Careless wiring can degrade the sound of any speaker system; even worse, it can damage your receiver. The first rule on installing speakers is to make sure the power is turned off on your receiver or amplifier while you are working with speaker connections.

KIND OF WIRE

Ordinary lamp cord with #18 gauge conductors is a good choice for most speaker systems. If you have long runs, more than 30 feet from receiver to speakers, get a cable with #16 gauge conductors, such as Radio Shack Cat. No. 278-1384.

If you must run a speaker cable under a rug, you can use flat television lead-in wire. Check the gauge of available twin-lead and get the one with the heaviest conductor, or lowest gauge number. Try to avoid long runs with this kind of wire.

TERMINALS

If your receiver has speaker terminals with an adjustable screw to clamp the conductor, it's a good idea to install spade lugs on your speaker cables. Choose a spade lug to match the size of screws on the receiver. If you don't have any spade lugs that fit, you can make a "j" hook at the end of the wire. Strip off about ½" of insulation from the end of the wire by cutting the insulation with a knife. If you slant the knife cut, as if you were sharpening a pencil, you will be less likely to cut the wire itself. To prevent a cable break, inspect the bared wire carefully for cut strands. This is especially important for cables inside the speaker enclosure where a failure will require major surgery to repair. Twist the strands together and tin the wire with rosin-core solder. This will prevent any splayed filaments that can short out your receiver. Form the soldered wire into a hook so it will wrap around the terminal screw without extending toward the adjacent terminal. For speaker systems with push-type quick hook-up terminals, prepare the cable the same way but without the hook.

Make sure the terminals are kept free of corrosion or any foreign material that can add resistance and degrade speaker performance.

HOW TO SOLDER

If a connection is to be soldered, use rosin-core solder. Never use acid-core with electronic equipment, and never let anyone use your electronics soldering tool with acid-core solder unless you plan to install a new tip.

Keep your soldering tools clean and tinned. A dirty tip conducts heat poorly, making bad joints. If the tip is dirty, use a piece of sandpaper to remove the metallic oxide; then heat the tip and rub solder over it until it is coated.

To solder a connection, first make a secure mechanical joint by threading the wire through the terminal and bending it tightly against the metal. Heat your iron, or gun, and hold the tip next to the joint. Stick some solder in the junction of the tip and the joint to improve heat conduction from tip to joint. Then apply the solder to the side of the joint opposite from the iron. Hold the iron and solder in position until the solder flows freely through the joint. When you remove the iron, the joint should be smooth and shiny.

In wiring your speakers, don't forget to observe polarity. If you fail to wire a pair of speakers correctly, bass performance and the stereo image will suffer. Or if you don't observe polarity when wiring the crossover network, there will be dips in the response curve. To test your system, refer to the tests in Chapter 9.

SPEAKER PLACEMENT IN THE ROOM

The most important rule on finding optimum room placement is to *experiment*. Sometimes even a minor change in position can make a big difference in listening quality.

Try to install the speakers with the drivers, most importantly the tweeters, near ear level. There is usually no problem in doing this with miniature shelf speakers or with large floor models, but with some in-between size speakers you must give more attention to proper placement. When such speakers are set on the floor against a wall, much of the treble is lost, and the bass often becomes too heavy. In such cases you can make a big improvement by putting the speakers on small stands. To find the optimum height, try stacking some large books or other support under the cabinets until the sound is right. Then buy or build a stand that will provide the required elevation. If you build one, make the front

Fig. 8-1. A speaker stand for enclosures of intermediate size raises and tilts the speaker, giving improved treble distribution.

edge slightly higher than the rear so the tweeters will be tipped upward. About 5 to 7 degrees is right for most rooms (Fig. 8-1).

Another precaution, make sure that the sound isn't blocked by bulky pieces of furniture. Again it is the treble that is most susceptible to problems of this kind; bass is much less directional.

The best places in your room for a good frequency response may not always be the best choice for the stereo image. As you know, for good stereo, the left and right channel speakers must have distance between them. The correct distance depends on the size of the room and how far you sit from the speakers. A plan of your listening room as seen from above should show the left and right channel speakers making an isosceles triangle with the listeners' chairs. The base of that triangle, the side with the speakers at each end, should not be the longest side. If furniture arrangement dictates a wide base, try angling the speakers inward, to fire at the listeners' chairs.

CABINET ORIENTATION

In the early days of stereo most compact speaker enclosures were made to be installed on tables or shelves with the long dimension horizontal. Today's speaker enclosures almost invariably are seen with the long dimension in the vertical plane (Fig. 8-2). This is not just a matter of fashion, but of good sound. No matter what kind of crossover network is used, there will be some overlap near the crossover frequency where the woofer and tweeter will produce the same frequency band. If there is a horizontal distance between two drivers that are producing the same frequencies, the path length of the sound from each driver to most listeners will be different (Fig. 8-3). This difference in path length causes phase shift which in turn introduces dips and peaks in the response curve.

Purists insist that the finest stereo reproduction can be achieved by keeping the line that passes through the woofer, mid-range and tweeter at a 90° angle to the line which extends from the speaker to the listeners. This is one reason for tilting low floor speakers upward. It follows from this rule that speakers placed near the ceiling should be aimed slightly downward.

PROTECTING YOUR SPEAKERS

If you are worried about blowing your speakers by driving them too hard, you can insert a fuse in series with each speaker system for protection. The fuse may alter the tonal response slightly by adding resistance. More resistance means a higher Q, or less damping on the woofer at resonance. The best way to decide whether this is good, bad, or irrelevant, is to try it.

To fuse the speakers, install a fuse holder on the back of the enclosure. Wire the fuse holder in series with one of the speaker leads. Table 8-1 shows some practical values for fuses to use with speakers of various power ratings.

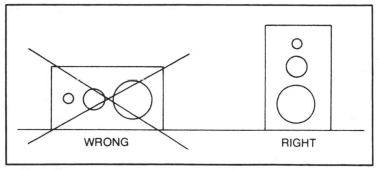

Fig. 8-2. Where possible, cabinet should be oriented so speakers form a vertical line.

If you have trouble with blown tweeters, the problem may be in your amplifier. Tweeters have more limited power handling ability than other drivers, but in the normal power distribution of music, they cause no trouble. An amplifier that clips when overdriven can destroy your tweeters even if the amplifier has a lower rated power output than the rated power of your speaker system. It does that by adding harmonics at frequencies that are multiples of the fundamental frequency. This adds power to the tweeter circuit at levels that didn't exist in the original signal. To protect your tweeters you can install Radio Shack's new dynamic protectors, Cat. No. 40-301. These protectors prevent damage to tweeters by amplifier spiking with virtually no effect on sound quality.

EXTENSION SPEAKERS

When you connect an extra set of speakers to your receiver and switch in both pairs, the second speakers appear in parallel with the main speakers. Assuming that each speaker is an 8-ohm speaker, the final impedance to the amplifier will be 4 ohms. These figures are only approximate because the impedance of all speakers vary with frequency. A problem arises when more than one 4-ohm speaker or a 4-ohm speaker and an 8-ohm speaker are operated

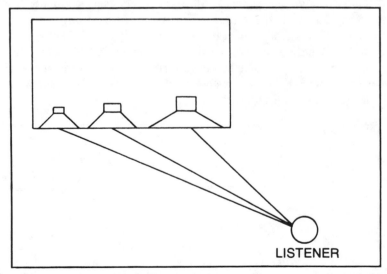

Fig. 8-3. When the line through the speakers in a cabinet is horizontal, the path length for the sound from each speaker to an off axis listener is different, causing phase distortion.

Table 8-1. Suggested Maximum Fuse Sizes for Various Speakers.

Speaker power rating	Maximum fuse rating. For better protection, use a fuse with ½ the current rating shown		
	4 ohm speakers	8 ohm speakers	16 ohm speakers
10	2	1	½
20	3	1½	¾
30	4	2	1
50	5	2½	1¼
75	6	3	1½

simultaneously in the same channel. Two 4-ohm speakers in parallel give a net impedance of 2 ohms. Or a 4-ohm speaker plus an 8-ohm speaker in parallel gives 2.7 ohms. Any value below about 3 ohms can permit too much current flow in the output stage of a solid state receiver or amplifier.

Here is the rule on impedance: if you plan to use more than a single pair of speakers at one time, make sure each speaker has an impedance of at least 8 ohms. If you plan to use only a single pair of stereo speakers, you can forget impedance.

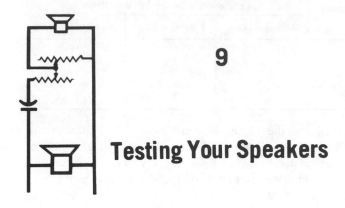

9

Testing Your Speakers

Speaker testing, as done by audio engineers, is a complicated business that requires expensive test equipment and even a special kind of fiberglass lined room, called a *dead room*. The engineers who design speakers use frequency response tests as a guide to monitor their work. For anyone else such tests aren't very useful.

Even if you have no test equipment, a few tests will help you get optimum performance from your speakers.

LISTENING TESTS

The most basic tests you can make, and in some ways the most useful ones, require little more than your two ears. Such tests do require that you identify subtle changes in sound that occur when you install speakers in different ways. Don't be discouraged if you fail to notice such changes at first. A trained ear requires practice.

FM HISS TEST

For this test you need an FM receiver and the speakers you want to test. To educate your ear, prepare a small reflector from a piece of hardboard or smooth particle board that is about 12″ × 12″ in size. Leave one side smooth, but glue or staple a 2″ to 4″ thick piece of fiberglass to the other side. Connect a bare speaker to one channel of the receiver, and, if possible, an enclosed speaker of the same model to the other channel. Turn the balance control all the way to the channel with the bare speaker. Disconnect the antenna from the receiver and tune for no signal, so that you get only the hiss that occurs between stations.

Hold the unbaffled speaker near your ear. Place the reflector behind the speaker with its smooth side toward the rear of the cone. Move the reflector in and out of the speaker's rear field as you listen for any changes in the sound. The reflector will probably produce a nasal quality by making the frequency response more peaky in the mid-range. Learn to identify this change of character when the reflector is in place.

Reverse the reflector so that the fiberglass pad faces the rear of the speaker cone. Repeat the procedure described above. Then reverse the reflector again, alternating from the smooth side to the side with the damping material. As you learn to recognize reflective coloration, you will appreciate the need for damping material behind enclosed cones.

Turn the balance control to favor the enclosed speaker, then back to the unbaffled one. If you hear a change in quality between the two speakers, it will be because of box coloration. Use this test to find the best amount and placement of damping material in the enclosure.

Engineers use a test similar to this to judge the degree of coloration in speakers. The only difference is that they use an expensive signal generator, one that produces a wide range of frequencies simultaneously. Such a generator is called a white noise generator and the test, a white noise test.

CRITICAL LISTENING

The only equipment you need here is your speakers, a tape deck or record player, and a receiver. Choose program material that you know; voice recordings are good for judging naturalness. If possible put your speakers outdoors on a quiet day to eliminate the effects of room acoustics. When listening to speakers indoors, even a slight change of position can produce a totally different room effect.

Listen first for a nasal quality, the kind of voice sound you get by holding your nose while speaking or singing. Of course you must discount any nasality that naturally occurs in any singer's voice. If the speaker fails this test, apply the FM hiss test and check the damping material in the box.

Next, adjust the volume control for a low sound level. If the speaker is peaky, you may hear only the peaks, making it sound thin or limited in range. This can be a rigorous test. It can be used indoors to help in adjusting mid-range and tweeter controls.

Don't be too dejected on the basis of a single program source. Commercial recordings and radio station transmissions vary greatly in audio quality. As you develop a critical listening ability, you will be amazed at the range in quality from such sources. And when a good source comes along, you'll be amazed at the kind of performance your speakers can deliver.

In some ways, this is the most important test you can make with your speakers. Speaker design should be based on a solid technical foundation, but there is an art to getting the best sound from a set of speakers.

TESTS WITH A SOUND LEVEL METER

One of the most helpful pieces of test equipment that you can own is a sound level meter. If you have ever struggled to make frequency response tests with microphones, cables, pre-amps, amplifiers, and voltmeters, you will immediately recognize the great convenience of being able to dispense with all these dangling cables and separate pieces of equipment and, instead, hold it all in the palm of your hand. But even though you can hold a sound level meter in your hand, you can get more reliable data by installing it on a camera tripod. Radio Shack's Realistic Sound Level Meter, Model No. 42-3019 is a perfect example of a piece of test equipment that combines convenient operation with low cost. With it you can test your entire stereo system, including the room as well as the components.

To use the sound level meter, mount it on a tripod and set the tripod a few feet from a speaker enclosure with the microphone end of the meter aimed at a 90° angle to a line between the meter and the speaker (Fig. 9-1). This position allows you to remove your body from the path of the sound where it would either block or reflect sound to the meter. The microphone pick-up pattern is non-directional at all but the highest frequencies.

Adjust the weighing selector to the C position. This produces a nearly uniform response curve from 32 to 8000 Hz. Set the response selector to the slow position. Put a test record on your phonograph. Most test records have an introductory test level tone of 1000 Hz; use this to adjust the range setting of the sound level meter so the meter will respond to the tone but isn't overdriven by it. Ordinarily you would find a suitable range band, say the 80 dB setting, then adjust the volume control on your amplifier or receiver so that the meter reads 0 dB. Make a record of the response level for every tone recorded on the test record, then plot it on a graph.

The next step is to correct any imbalance you find on your graph or data table. Minor variations in response levels can often be corrected by use of the mid-range and tweeter controls. If your data shows a huge peak in bass response, try pulling the speaker cabinet out from the wall a few inches. Or set it up on a pile of books and test again. When you find the optimum position, you can buy or make a permanent stand for it.

You can also use your sound level meter to test the dispersion characteristics of your speakers. Set up the sound level meter at your usual listening position and check to see that you are getting good high frequency coverage there. Note that you must make allowance for the normal reduced sensitivity of the microphone in the sound level meter to high frequency sound from the 90° angle at which you will use it. A polar graph in the booklet that comes with the meter shows that the response is down about 5 dB for 8000 Hz at 90°, and about 7½ dB for 15,000 Hz at 90°.

If you have a compact speaker that sits on the floor, you may find that a speaker stand will improve dispersion, especially if the stand allows for a slightly angled tilt. The sound level meter will help you get the correct angle.

FREQUENCY CORRECTION WITH AN EQUALIZER

For an easy way to correct the frequency imbalances you may have discovered with your sound level meter, get a graphic equalizer. A low-cost model, such as Radio Shack Model No. 31-1988, will correct many of the problems caused by imperfect room acoustics. Or, for a more versatile control over even nar-

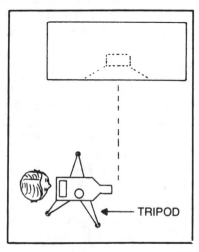

Fig. 9-1. Test set-up with sound level meter.

TRIPOD

rower frequency bands, you can go all out for the 10-band model, Model No. 31-2000.

Graphic frequency equalizers are the last step in the evolution of tone controls which began with a single knob on old radios that cut the highs. Next, hi-fi amplifiers boasted dual bass and treble controls that permitted boost as well as cut functions. These were not always adequate because tweeters usually have higher efficiency than woofers, and using the amplifier tone control to cut the tweeter puts the upper highs below audibility, so mid-range and tweeter controls were introduced. Now, with graphic equalizers, it is possible to focus on the particular frequency range that is too weak, or too prominent, and correct that band without disturbing the overall tonal balance. This ability not only corrects for room acoustics, but it can make a low-priced speaker sound like a much more expensive model.

POLARITY TESTS

For the reasons mentioned in Chapter 1, it is essential that your stereo speakers be connected so that they are in phase with each other. In the tests that follow, notice that you must alter the connections on one speaker only. If you switch the leads to both speakers, their polarity with respect to each other is the same as in the first connection.

Listening Test

Place the two speakers close together, face to face. Feed a 100 Hz tone from a test record to them, or select music with heavy bass tones. Reverse the leads to one speaker. The correct hook-up is the one with the greater bass response.

Battery Test

Sometimes it isn't convenient to move the speakers close together, particularly if the speakers are large. You can use an ordinary flashlight battery to confirm the speakers' polarity.

Connect the positive pole of the battery to one speaker box terminal and the negative pole to the other while someone watches the woofer cone. Reverse the connections if necessary to find the connection where the cone moves forward, or outward from the box, when the connection is made, backward when broken. When you find this connection, mark the speaker terminal to which the positive battery pole is connected with a red dot. Do this for each speaker and observe these marks when you connect the speakers to your receiver.

If you can't remove the grille cloth, you can make this test by holding a single sheet of newspaper over the grille cloth in front of the woofer. Watch the movement of the newspaper to determine the direction of woofer cone travel.

As mentioned, you need nothing more than a battery and some wires to make this test, but the speaker damper tester, described later in this chapter, provides a more convenient way of hooking up a battery without loose connections that can confuse the results.

DAMPING TEST

To test your speaker and enclosure combinations for proper damping, you need only a cheap homemade tester. The circuit of the tester is shown in Fig. 9-2 and a list of parts in Table 9-1. This tester can be built in an empty tuna can. The resistor can be any value from ¼ to ½ ohm. Ideally it should be based on the damping factor of the receiver or amplifier that you will use with the speaker, but the value range given above will work well with modern receivers.

Connect the leads from the tester to your speaker. Flip the switch on and off, listening to the sound of the woofer at each flip. If the damping is inadequate, you will hear a thud or a bong instead of a click. If the sound is a bong on the "make" connection as well as on the "break," the speaker is seriously underdamped. If the sound is a click when you make connection, but a bong on the break, it is slightly underdamped. If it is click on each connection, the speaker is adequately damped, perhaps even overdamped. If the speaker is overdamped, the box may be too large. You can reduce cubic volume by adding bricks or blocks of wood.

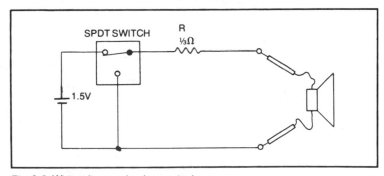

Fig. 9-2. Wiring diagram for damper tester.

Table 9-1. Parts List for Damper Tester.

1	SPDT Mini Toggle Switch	Radio Shack Cat. No. 275-326
1 pkg.	Insulated Alligator Clips	Radio Shack Cat. No. 270-378
1	Test Probe Wire, Red	
1	Test Probe Wire, Black	
1	Battery Holder, to fit size C, 1.5 Volt Battery	
1	Small Chassis (use tuna can)	
2	Banana Plugs and Jacks (if detachable test leads are desired)	

To add damping, staple a blanket of fiberglass over the back of the speaker frame. Stretch the fiberglass as you staple it, so that it is taut rather than floppy. A thickness of about 1″ is usually adequate.

This test is particularly useful for speakers in ported boxes because such systems require heavier damping than closed box speakers to avoid the boom box effect. It is difficult to measure the damping of ported box speakers any other way without expensive test equipment.

Make the final adjustment of damping material by a listening test.

10

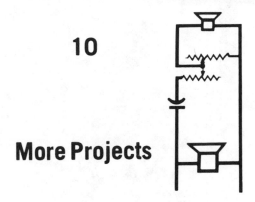

More Projects

If you do any of these projects, you can alter the construction methods, but try to make the internal cubic volume of the box equal to that of the plans. Refer to earlier projects for details of any particular construction method and to Chapter 3 for guidelines on desirable box qualities.

PROJECT 10: TWO-WAY IN A PINE BOX

Here is a low-cost 2-way system that offers smooth sound (Fig. 10-1). Pine boards are adequate for small enclosures, but here the pine is lined with ½" particle board for extra strength and rigidity (Fig. 10-2 and Table 10-1). Build the pine shell first, then glue in the liner. The liner should be recessed 1⅛" from the front edges and ¾" from the back edges of the box walls. The speaker board and back panel can be glued and nailed directly to the liner, no cleats needed. The internal dimensions should be about 9½" × 17½" × 7⅜".

The 16 μF capacitor that comes with the mid-range tweeter can be used as a high-pass filter (Fig. 10-3). An 8-ohm L-pad to compensate for varying acoustical conditions completes the project.

PROJECT 11: EXTENSION SPEAKER

This extension speaker can be built without power tools. By painting the box flat black and using a storage crate to cover and protect it, minor flaws in your cabinet work will not be noticeable (Fig. 10-4).

For a local volume control on a stereo pair of these speakers,

Fig. 10-1. Pine box and 2-way system combine to give smooth low-cost sound.

you can install a double L-pad in a standard wall box of the room where the speakers are installed (Table 10-2). Or you can install a single L-pad in the back of each speaker (Fig. 10-7).

For easy wall mounting, get a set of metal wall brackets and

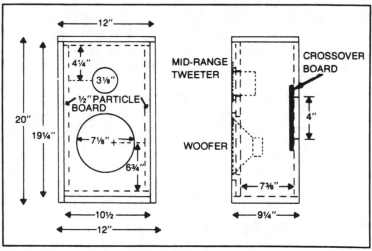

Fig. 10-2. Enclosure plans for Project 10.

142

Table 10-1. Parts List for Project 10.

1″ × 10″ Pine (real dimensions ¾″ × 9¼″):	
2 9¼″ × 19¼″	Sides
1 9¼″ × 12″	Top
1 9¼″ × 10½″	Bottom
½″ Particle Board:	
2 7⅜″ × 10½″	Top & bottom liner
2 7⅜″ × 17½″	Side liner - cut to fit
¾″ Particle Board:	
2 10½″ × 18½″	Speaker board & back
Speakers and Components:	
1 8″ woofer	Radio Shack Cat. No. 40-1006
1 Mid-range tweeter	Radio Shack Cat. No. 40-1289
1 8-ohm L-pad	Radio Shack Cat. No. 40-980
Miscellaneous:	
Fiberglass insulation	Radio Shack Cat. No. 42-1082

install one in the upper back of each speaker (Fig. 10-6). If you prefer a more conventional looking extension speaker, use the enclosure plans for Project 10 for the 8″ full-range speaker.

PROJECT 12: LARGE WOOFER TWO-WAY SPEAKER

The introduction of a new Radio Shack mid-range tweeter that can be used down to 700 Hz makes this 2-way system practical. The 10″ woofer offers deeper bass than is possible in most 2-way systems (Fig. 10-8).

Note that the wiring circuit for this project is identical with that of Project 10, each one using the 16 μF high-filter capacitor supplied with the mid-range tweeter (Fig. 10-3).

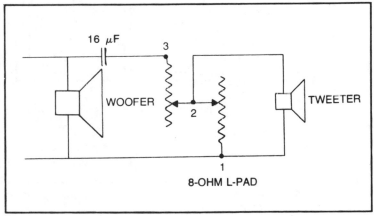

Fig. 10-3. Wiring diagram for Projects 10 and 12.

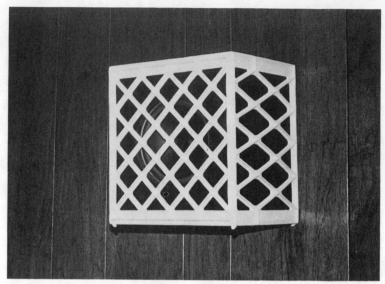

Fig. 10-4. Storage crate protects and enhances appearance of extension speaker.

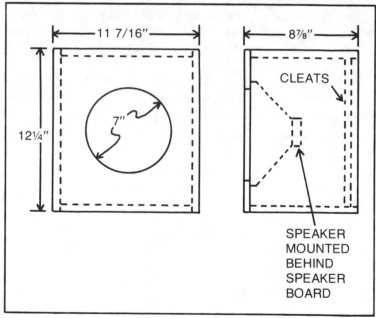

Fig. 10-5. Enclosure plans for Project 11. If a more conventional box is desired, build one to the plans for Project 10.

144

Fig. 10-6. Metal bracket in upper back makes wall installation easy.

The enclosure shown was built from birch plywood (Fig. 10-8 and Table 10-3). The walls were assembled with beveled corner joints, and cleats were installed to hold the speaker board and back panel. The enclosure for Project 2 has the same internal volume and can be used if you change the speaker cut-outs (Fig. 10-9).

PROJECT 13: A PORTED SYSTEM WITH SPECIAL DAMPING

In testing a 6½" woofer, Radio Shack Model No. 40-1009, I found different values from the parameters published in the catalog. My tests showed:

$$f_s = 57.5 \text{ Hz}$$
$$Q = 0.6$$
$$V_{AS} = 0.7 \text{ cu. ft.}$$

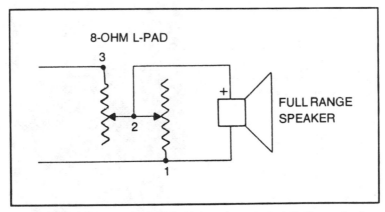

Fig. 10-7. Wiring diagram for Project 11. L-pad may be installed in speaker box or, if stereo control is used, in wall switch box.

½" Plywood
1	117/16" × 12¼"	Speaker board
2	8⅜" × 12¼"	Sides
2	8⅜" × 10½"	Top & Bottom
1	10¼" × 11⅛"	Back

Crate:
1	13¼" × 12¼" × 9½"	Sterilite No. 1590

Speaker and Components:
1	8" Full range speaker	Radio Shack Cat. No. 40-1286
1	Terminal plate	Radio Shack Cat. No. 274-625
1	Stereo volume control	Radio Shack Cat. No. 40-978
	Or use 1 L-pad in each enclosure	

Miscellaneous:
Fiberglass insulation	Radio Shack Cat. No. 42-1082
Metal Wall bracket	Radio Shack Cat. No. 40-150

The optimum volume ported box turned out to be 2.4 cu. ft., rather large for a 6½" woofer. However with a 1" layer of fiberglass stapled tightly over the back of the woofer (Fig. 10-12), the Q can be lowered to 0.5. This reduces the optimum volume to 1.43 cu. ft. Because this woofer is small for a box of such volume, less

Fig. 10-8. Large 10" woofer gives this speaker deeper bass than is possible with most 2-way systems.

Table 10-3. Parts List for Project 12.

¾″ Birch Plywood:	
2 10⅞″ × 15½″	Top & bottom
2 10⅞″ × 23½″	Sides
Strip Veneer, ¾″ wide:	
7 feet	To cover front edge of plywood
¾″ Particle Board:	
2 14″ × 22″	Speaker board & back
¾″ Pine:	
18 feet ¾″ × ¾″	Cleats and corner blocks
¼″ Plywood:	
1 13⅞″ × 21⅞″	Grille board
Speakers and Components:	
1 10″ woofer	Radio Shack Cat. No. 40-1331B
1 Mid-range tweeter	Radio Shack Cat. No. 40-1289
1 8-ohm L-pad (optional)	Radio Shack Cat. No. 40-977
Miscellanous:	
Fiberglass	Radio Shack Cat. No. 42-1082
Grille cloth	

Note: If desired, the enclosure for Project 2 may be used.

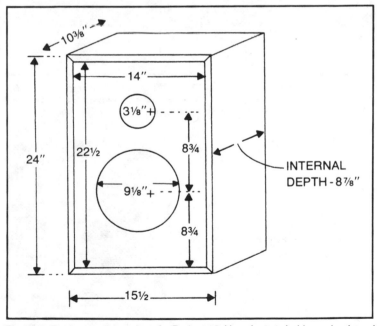

Fig. 10-9. Enclosure dimensions for Project 12. Use cleats to hold speaker board and back. Internal volume is identical to that of box for Project 2. Project 2 box can be used for Project 12 with only the speaker board cut-outs changed.

147

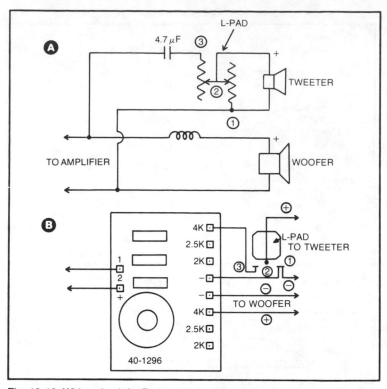

Fig. 10-10. Wiring circuit for Project 13. A is the schematic diagram of the 4000 Hz crossover, and B shows how to wire 2-way crossover network, Radio Shack Model No. 40-1296. Don't use the capacitor mounted on the tweeter.

Fig. 10-11. Project 13, complete except for grille cloth.

148

Table 10-4. Parts List for Project 13.

¾" Plywood:
2   10½" × 24"   Sides
2   10¼" × 15"   Top & bottom
¾" Plywood or Particle Board:
2   13½" × 22½"   Speaker board & back
Pine:
4   ¾" × ¾" × 13½"   Cleats
4   ¾" × ¾" × 21"   Cleats
4   ¾" × ¾" × 7¼"   Corner glue blocks
Grille Frame:
As desired
Paperboard Tube:
1   2⅜" Inside diameter × 1¾"   Port
      Or make a port with square cross section, 2 7/16" × 2 7/16"
Speakers and Components:
1   6½" Woofer   Radio Shack Cat. No. 40-1009
1   ¾" Hard dome tweeter   Radio Shack Cat. No. 40-1376
1   2-way Crossover network   Radio Shack Cat. No. 40-1296
1   8 Ohm L-pad   Radio Shack Cat. No. 40-977
1   Terminal plate   Radio Shack Cat. No. 274-625
Miscellaneous:
Fiberglass insulation   Radio Shack Cat. No. 42-1082
Grille cloth   As desired

than 10% overvolume was planned, making a box of 1.5 cu. ft. a practical choice (Fig. 10-11).

In wiring this, don't use the capacitor supplied with the tweeter.

You'll find this system, with the special damping, gives a controlled natural sounding bass response. If you like a more dominant bass, you can try it without the damping pad on the woofer.

PROJECT 14: CAR SPEAKER

Use ¼" tempered hardboard for this little box (Fig. 10-14). If you have power tools, dado the side edges of the top and bottom and all around the speaker board, as shown in the plans (Fig. 10-15). If you have no power tools, use butt joints.

Glue the box walls and speaker board together. Use the hardware that comes with the moulded speaker grille (Table 10-5) to install the grille and the 4" speaker. If you don't plan to use the grille, you can install the 4" speaker from outside the box, but it will be unprotected. No crossover network is required (Fig. 10-16).

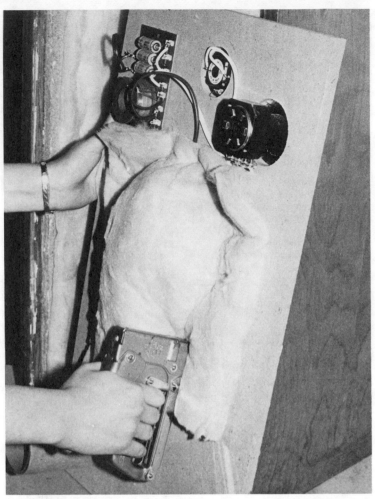

Fig. 10-12. Staple a 12" × 12" sheet of 1" fiberglass over the woofer. Fiberglass pad should be pulled taut.

Secure the speaker to the shelf under the back car window by locating bolts in the box to match the locations of speaker mounting bolts in the car body. If you can find bolts that are long enough, you can run the mounting bolts from the speaker board through the box and shelf. For this kind of installation the shelf serves as the back of the enclosure. Fill the space under the box with loose damping material. If you can't find long bolts, use shorter bolts through the back and glue the back to the box.

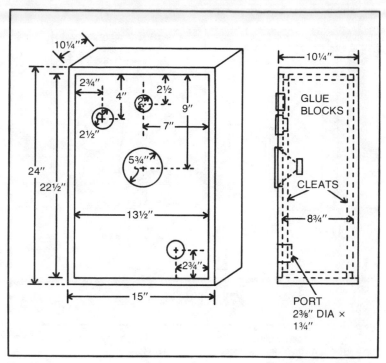

Fig. 10-13. Enclosure plans for Project 13.

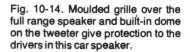

Fig. 10-14. Moulded grille over the full range speaker and built-in dome on the tweeter give protection to the drivers in this car speaker.

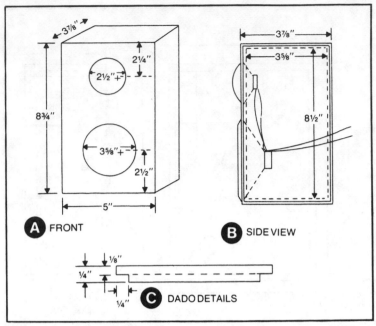

Fig. 10-15. Enclosure plans for Project 14. Dado is used on each side edge of top and bottom, on all edges of front and back panels.

PROJECT 15: UTILITY MUSICAL INSTRUMENT SPEAKER

This project is for those musicians who like the sound of an open-backed speaker. It gives a more pronounced middle bass response, but the user should be aware that the power input to the speaker must be limited somewhat more than if it were totally enclosed. This disadvantage isn't as serious as it sounds; the open back enclosure gives a middle bass resonance that makes the system more efficient than it would be with a solid back.

Table 10-5. Parts List for Project 14.

¼" Hardboard:		
2	3⅝" × 8½"	Sides
2	3⅝" × 5"	Top & bottom
2	5" × 8¾"	Speaker board & back
Speakers and Components:		
1	4" full-range speaker	Radio Shack Cat. No. 40-1197
1	Piezo speaker with dome	Radio Shack Cat. No. 40-1382
1	5" moulded speaker grille	Radio Shack Cat. No. 40-1291
Miscellaneous:		
	Fiberglass insulation	Radio Shack Cat. No. 42-1082
	Speaker cable	

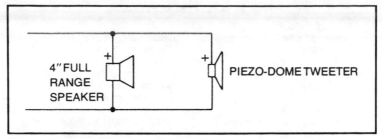

Fig. 10-16. Wiring diagram for Project 14.

Cut out the parts and assemble the shell with glue and nails (Fig. 10-17 and Table 10-6). Install cleats, recessed 1¼" from the front edges, ¾" from the rear edges of the walls. Glue and nail down the speaker board and the rear short panel. Cover the box with cloth-backed vinyl, following the instructions for Project 8. Install metal corners, then the speakers (Fig. 10-18). Make up a grille board and cover it with grille cloth. Install it with ornamental screws.

If desired, you can install a carrying handle on each side.

PROJECT 16: PA SPEAKER

This column speaker (Fig. 10-19) gives excellent voice reproduction, but its bass response is more limited than that of most stereo or musical instrument speakers. The vertical column provides good horizontal sound distribution with good freedom from feedback (Fig. 10-20).

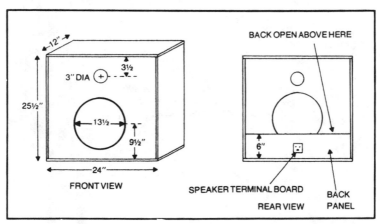

Fig. 10-17. Enclosure plans for Project 15. Cover enclosure with cloth-backed vinyl and use metal corners, as was done in Project 8.

Table 10-6. Parts List for Project 15.

¾" Particle Board:

4	12" × 24"	Sides
1	22½" × 24"	Speaker board
1	6" × 22½"	Rear panel

¾" Pine:

12 feet ¾" × ¾"	Cleats for speaker board and rear panel

Speakers and Components:

1	15" musical instrument speaker	Radio Shack Cat. No. 40-1315
1	Piezo super horn	Radio Shack Cat. No. 40-1381
1	Speaker terminal board	Radio Shack Cat. No. 274-624

Miscellaneous:

Cloth-backed vinyl
Metal corners
Grille cloth

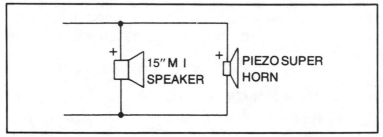

Fig. 10-18. Wiring diagram for Project 15.

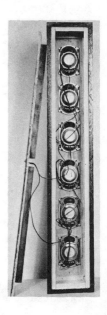

Fig. 10-19. This column speaker gives good voice reproduction and improved horizontal distribution.

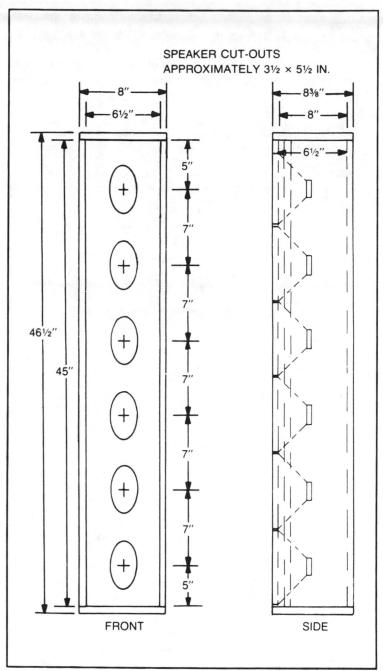

SPEAKER CUT-OUTS
APPROXIMATELY 3½ × 5½ IN.

FRONT

SIDE

Fig. 10-20. Enclosure plans for Project 16.

If you install the speakers behind the speaker board, round off the front edges of the cut-outs. Loosely fill the enclosure with cut pieces of damping material.

For a 12-ohm impedance, wire the speakers in 3 paralleled pairs, then wire the 3 pairs in series. Or, for an impedance of about 5 ohms, put 2 groups of 3 speakers in parallel, then wire the 2 groups in series (Fig. 10-21).

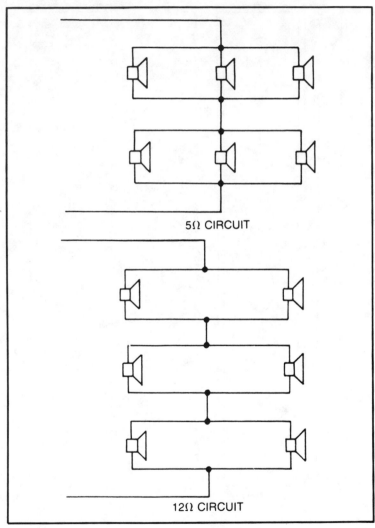

Fig. 10-21. Choose the wiring circuit to match the impedance you need for Project 16.

Table 10-7. Parts List for Project 16.

¾" Plywood:		
	2 8⅜" × 45"	Sides
	2 8⅜" × 8"	Top & bottom
	2 6½" × 45"	Front & back
¾" Pine:		
	18 feet ¾" × ¾"	Cut into cleats for front
		and back
¼" Plywood:		
	1 6⅜" × 44⅞"	Grille board
Speakers:		
	6 4" × 6"	Radio Shack Cat No. 40-1298
Miscellaneous:		
	Fiberglass	Radio Shack Cat. No. 42-1082
	16-gauge speaker wire	Radio Shack Cat. No. 278-1384
	Grill cloth	

With the speakers specified (Table 10-7), the rated power handling capacity is 180 watts. For small rooms you can shorten the column and use four speakers. Wire them in two paralleled pairs, the pairs in series for a standard 8-ohm impedance.

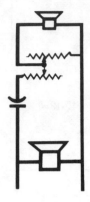

Appendix A
Useful Formulas

ELECTRICAL UNITS USED IN SPEAKER CIRCUITS

Inductance

The unit of inductance is the henry.

All smaller units should be converted to henrys before using the formulas.

$$1 \text{ millihenry (mH)} = 0.001H \text{ or } 1 \times 10^{-3}H$$
$$1 \text{ microhenry } (\mu H) = 0.000001H \text{ or } 1 \times 10^{-6}H$$

Capacitance

The unit of capacitance is the farad.

All smaller units should be converted to farads before using the formulas.

$$1 \text{ microfarad } (\mu F) = 0.000001 \text{ farad or } 1 \times 10^{-6} \text{ farad}$$
$$1 \text{ picofarad (pF)} = 0.000000000001 \text{ farad or } 1 \times 10^{-12} \text{ farad}$$

OHM'S LAW FOR SPEAKERS

For alternating current, the Ohms's Law formulas are:

$$E = IZ$$
$$I = E/Z$$
$$Z = E/I$$

E is the voltage in volts,
I is the current in amperes and
Z is the impedance in ohms.

Worked Examples:

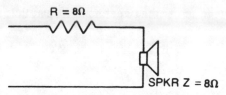

An 8Ω resistor is put in series with an 8Ω tweeter.
How much is the sound level reduced?
Assume any drive voltage that is convenient; for example, 16 V.
The nature of the speaker's impedance depends on the frequency,
so we will assume that it is purely resistive.

Then:

$$I = 16/16 = 1\ A$$

The voltage across the speaker is:

$$E = 1 \times 8 = 8\ V$$

Before adding the resistor, the voltage across the speaker was
16 V. So the ratio of voltage in the new circuit to the old one is:

$$\frac{\text{New voltage}}{\text{Old voltage}} = \frac{8}{16} = 0.5$$

The dB/voltage ratio chart tells us that for a voltage ratio of
0.5, the output is down 6 dB.

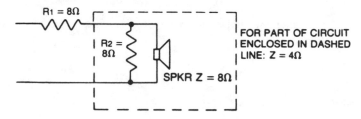

FOR PART OF CIRCUIT
ENCLOSED IN DASHED
LINE: Z = 4Ω

In the same tweeter circuit, an additional resistor is added
parallel to the tweeter. What is the tweeter output now compared
to its output with no resistors in its circuit? First, get the
impedance of the tweeter and its parallel resistor. Since the
paralleled resistances are equal, then the net impedance of the
tweeter and paralleled resistor branch is 8/2 or 4Ω.
The total impedance in the circuit is $8 + 4 = 12Ω$.
And:

$$I = 16/12 = 1.33\ A$$

The voltage across the tweeter and its parallel resistor branch is:

$$E = 1.33 \times 4 = 5.32\,V$$

And the tweeter's output is:

$$\frac{\text{New voltage}}{\text{Old voltage}} = \frac{5.32}{16} = 0.33$$

From the dB chart, the output is down 10 dB.

REACTANCE & RESONANCE

Capacitors

Capacitive reactance:

$$X_c = \frac{1}{2\pi fC}$$

To get the value of a capacitor that will give a specified reactance at a given frequency:

$$C = \frac{1}{2\pi fX_c}$$

Inductance

Inductive reactance:

$$X_L = = 2\pi fL$$

To get the value of a choke that will give a specified reactance at a given frequency:

$$L = \frac{X_L}{2\pi f}$$

Resonance

$$f = \frac{1}{2\pi\sqrt{LC}}$$

Worked example for resonance:
A 1.875 mH choke is wired into a peak suppressor filter circuit with a 7.5 μF capacitor. What frequency will be most affected?

$$f = \frac{1}{6.28\sqrt{0.001875 \times 0.0000075}} = 1342\,Hz$$

Notice that in the formula for resonance, it is the *product* of L × C that determines the frequency of resonance. You can increase L and decrease C by the proper amounts and still have the same frequency. This permits you to make filters that are effective at the frequency you want with various values of components.

For parallel resistors:

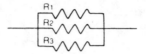

Where $R_1 = R_2 = R_3$, and N is the number of resistors

$$\text{Total } (R_T) = R/N$$

For mixed value resistors:

$$R_T = \frac{1}{1/R_1 + 1/R_2 + 1/R_3}$$

If n = 2

$$R_T = \frac{R_1 \times R_2}{R_1 + R_2}$$

For series resistors:

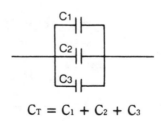

$$R_T = R_1 + R_2 + R_3$$

For parallel capacitors:

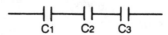

$$C_T = C_1 + C_2 + C_3$$

For series capacitors:

Where $C_1 = C_2 = C_3$ and N is the number of capacitors.

$$C_T = C/N$$

For mixed value capacitors:

$$C_T = \frac{1}{1/C_1 + 1/C_2 + 1/C_3}$$

If N = 2:

$$C_T = \frac{C_1 \times C_2}{C_1 + C_2}$$

For parallel chokes:

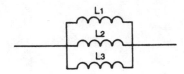

Where $L_1 = L_2 = L_3$, and N is the number of chokes.

$$L_T = L/N$$

For mixed value chokes:

$$L_T = \frac{1}{1/L_1 + 1/L_2 + 1/L_3}$$

If N = 2:

$$L_T = \frac{L_1 \times L_2}{L_1 + L_2}$$

For series chokes:

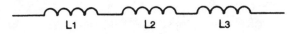

$$L_T = L_1 + L_2 + L_3$$

For resistor and capacitor in series:

R   C

$$Z = \sqrt{R^2 + X_c^2}$$

Worked example:
Let R = 10 Ω, C = 0.000002 or 2 μF.
What is the impedance at 10,000 Hz?

$$X_c = \frac{1}{2\pi f C} = \frac{1}{(6.28)\ (10,000)\ (0.000002)} = 8\Omega$$

And:
$$Z = \sqrt{100 + 64} = 12.8\,\Omega$$

For resistor and capacitor in parallel:

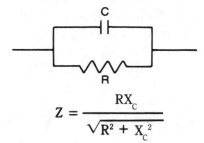

$$Z = \frac{RX_C}{\sqrt{R^2 + X_C^2}}$$

Worked example:

Let $R = 10\,\Omega$, $C = 0.000002$ f or $2\,\mu$F.

What is the impedance at 10,000 Hz?

From the example above: $X_C = 8\,\Omega$ at 10,000 Hz.

So:

$$Z = \sqrt{\frac{10 \times 8}{100 + 64}} = \frac{80}{12.8} = 6.25\,\Omega$$

For resistor and choke in series:

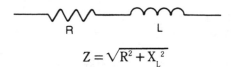

$$Z = \sqrt{R^2 + X_L^2}$$

Worked example:

Let $R = 10\,\Omega$, $L = 0.000127$ H or 0.127 mH.

What is the impedance at 10,000 Hz?

$$X_L = 2\pi fL = 6.28 \times 10{,}000 \times 0.000127 = 8\,\Omega$$

So:

$$Z = \sqrt{100 + 64} = 12.8\,\Omega$$

For resistor and choke in parallel:

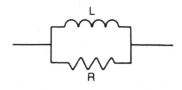

$$Z = \frac{RX_1}{\sqrt{R^2 + X_L^2}}$$

Worked example:

Let $R = 10\Omega$, $L = 0.000127$ H or 0.127 mH.

What is the impedance at 10,000Hz?

From the example above:

$$X_L = 8\Omega \text{ at } 10,000 \text{ Hz.}$$

So:

$$Z = \frac{10 \times 8}{\sqrt{100 + 64}} = \frac{80}{12.8} = 6.25\Omega$$

For a resistor, choke, and capacitor in parallel:

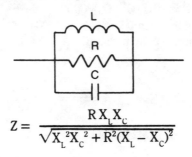

$$Z = \frac{RX_LX_C}{\sqrt{X_L^2X_C^2 + R^2(X_L - X_C)^2}}$$

Worked example:

Let $R = 10\Omega$, $L = 0.00159$ or 1.59 mH, and $C = 0.000016$ or 16 μF.

What is the impedance at 1000 Hz?

$$X_L = 2\pi fL = (6.28) \times 1000 \times 0.0015 \approx 10\Omega$$

$$X_C = \frac{1}{2\pi fC}$$

$$= \frac{1}{6.28 \times 1000 \times 0.000016} \approx 10\Omega$$

$$Z = \sqrt{\frac{10 \times 10 \times 10}{100 \times 100 + 100(10 - 10)^2}}$$

$$= \frac{1000}{100} \approx 10\Omega$$

What is the impedance at 500 Hz?

At 500 Hz the reactance of L will be halved, compared to 1000 Hz, and that of C will be doubled.

So: $X_L = 5\Omega$ and $X_c = 20\Omega$ at 500 Hz.

$$Z = \sqrt{\dfrac{10 \times 5 \times 20}{25 \times 400 + 100 (5 - 20)^2}}$$
$$= \sqrt{\dfrac{1000}{32500}} = \dfrac{1000}{180.3} = 5.55\,\Omega$$

CONVERSION FACTORS FOR SPEAKER COMPLIANCE

Speaker compliance is sometimes measured in cm/dyne- the CGS (centimeter-gram-second) system because the centimeter is a convenient unit to use in measuring cone diameter. Sometimes compliance is shown in mixed units of mm/Newton or even as its reciprocal term—stiffness—in Newton/Meter, the MKS (meter-kilogram-second) system.

To change cm/dyne to mm/Newton:

$$1\,cm = 10mm$$
$$1 \times 10^5\,dynes = 1\,Newton$$

So:

$$\dfrac{1\,cm}{dyne} \times \dfrac{10\,mm}{cm} \times \dfrac{1 \times 10^5\,dynes}{Newton}$$

cancels out to:

$$1 \times 10\,mm \times \dfrac{1 \times 10^5}{Newton} = 1 \times 10^6\,mm/Newton$$

So:

$$C_{ms}\,(cm/dynes) \times 1 \times 10^6 = C_{ms}\,(mm/N)$$

To change C_{ms} in cm/dynes to stiffness (S) in Newton/Meter, first change C_{mx} in cm/dynes to S in dynes/cm by:

$$S = 1/C_{ms}$$

And:

$$\dfrac{1\,dyne}{cm} \qquad \dfrac{1\,Newton}{1 \times 10^5\,dynes} \times \dfrac{1 \times 10^2\,cm}{1\,meter}$$

cancels out to:

$$1 \times \dfrac{1\,Newton}{1 \times 10^5} \times \dfrac{1 \times 10^2}{1\,meter} = 1 \times 10^3\,Newton/meter$$

APPROXIMATE VOLTAGE RATIO (+)	dB	APPROXIMATE VOLTAGE RATIO (−)
1	+0−	1
1.1	1	0.9
1.25	2	0.8
1.4	3	0.7
1.6	4	0.6
1.8	5	0.55
2	6	0.5
2.25	7	0.45
2.5	8	0.4
2.8	9	0.35
3.2	10	0.32
3.5	11	0.28
4	12	0.25
5.6	15	0.18
8	18	0.13
16	24	0.06

TO CALCULATE ANY VOLTAGE RATIO:

$$dB = 20 \, \text{LOG} \, E1/E2$$

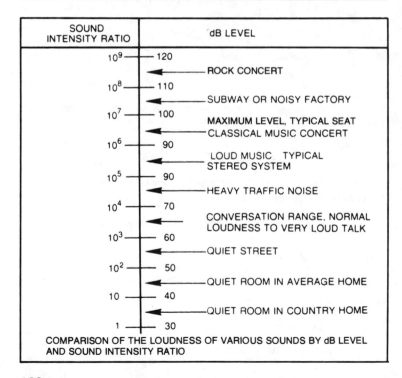

COMPARISON OF THE LOUDNESS OF VARIOUS SOUNDS BY dB LEVEL AND SOUND INTENSITY RATIO

So: S in dynes/cm × 1 × 10^{-3} = S (Newton/Meter)

Worked example:

The compliance of 12 in. speaker is listed as 0.416 × 10^{-6} cm/dynes.

What is the stiffness in N/m?

$$S(dynes/cm) = \frac{1}{0.417 \times 10^6} = 2.4 \times 10^6 \text{ dynes/cm}$$

$$S (N/m) = 2.4 \times 10^6 \text{ dynes/cm} \times 1 \times 10^{-3}$$

$$= 2.4 \times 10^3 \text{ N/m}$$

$$= 2400 \text{ N/m}$$

SPEAKER FORMULAS FOR CONSTANT VOLTAGE SOUND DISTRIBUTION SYSTEMS

Power formulas:

$$P = \frac{E^2}{Z} \qquad Z = \frac{E^2}{P}$$

where:

P = the power in watts,

E = the voltage in volts and

Z = the impedance in ohms.

TO CHOOSE IMPEDANCE MATCHING TRANSFORMERS:

1. Decide how much power is to be available to each speaker by:

a. Cubic volume and acoustical character of each location.

b. Desired sound level at each location.

c. Available amplifier power—make sure total isn't exceeded.

2. Substitute the transformer primary impedance for Z and the voltage of distribution system for E in the formulas above to find the right impedance transformer.

Worked example: A 50 W amplifier is used in a 70 V line. What impedance transformer is needed where the desired power to the speaker is 10 W.

$$Z \text{ (of transformer)} = \frac{(70)^2}{10} = \frac{5000}{10} = 500\Omega$$

The needed transformer should have a primary impedance of 500Ω and a secondary impedance to match that of the speaker. No more than five such speakers can be used with this amplifier.

To choose the correct constant voltage transformer tap:

1. Choose a transformer with a matching secondary impedance—an 8 Ω secondary for an 8Ω speaker.

2. Connect the desired power tap—10W for above example—to speaker.

3. Connect the constant voltage distribution line to the primary.

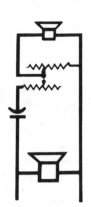

Appendix B
Tests Not
Covered in Chapter 9

These tests require a sine wave generator and an AC voltmeter. The test set-up is shown in Fig. B-1.

IMPEDANCE

Get a 10-ohm precision resistor, 1%, ½-watt or greater, for R_2 in Fig. B-1. Connect it at points x and y. Adjust generator output to read any 10 units on voltmeter, but choose a point that uses about 10% of the scale. For example, with a 1-volt scale, the reading should be 0.1 volt. After calibrating the generator output with this resistor, the voltage scale can be read in ohms, 0.1 volt = 10 ohms, 0.2 volts = 20 ohms, and so on.

Connect the speaker to points x and y. Run an impedance curve, recording values at each 10 Hz interval to 100 Hz, every 100 Hz to 1000 Hz, and at each 1000 Hz interval above that. If your speaker's impedance rises enough to require a higher voltage scale, you can either recalibrate the generator so that 0.05 V = 10 ohms, or get another precision resistor of more than 100 ohms and calibrate to a higher voltage scale.

FREE AIR RESONANCE

Suspend the speaker in mid-air with the cone in its normal vertical position. Vary the frequency from the sine wave generator until you find the low frequency that produces the greatest rise in impedance. Record this frequency as f_s.

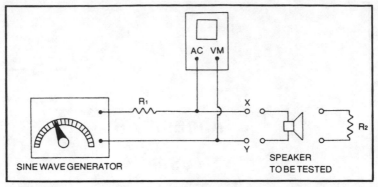

Fig. B-1. Test set-up to measure speaker impedance, free air resonance, and Q. $R_1 = 200$ to 1000 ohms.

DRIVER Q

Calibrate an ohm-meter with a 5 to 10 ohm precision resistor. Measure the DC resistance of the voice coil. Record this value as R_e.

Set the sine wave generator to the frequency of f_s. Read the impedance value and record it as Z_{max}. Calculate the value of r_o from:

$$r_o = \frac{Z_{max}}{R_e}$$

Record this value.

Find $\sqrt{r_o}$ and record its value.

Calculate the value of reduced impedance (Z') where:

$$Z' = \sqrt{r_o} \times R_e$$

Find the frequencies below and above f_s where driver impedance equals Z'. Record these frequencies as f_1 and f_2.

Check the accuracy of your work by this test:

$$f_s = \sqrt{f_1 \times f_2}$$

The solution to this formula should be accurate within about 1 Hz or 2%, whichever is greater.

Find the speaker's mechanical Q (Q_{ms}) by:

$$Q_{ms} = \frac{f_s \sqrt{r_o}}{f_1 - f_2}$$

And its electrical Q (Q_{es}) by:

$$Q_{es} = \frac{Q_{ms}}{r_o - 1}$$

170

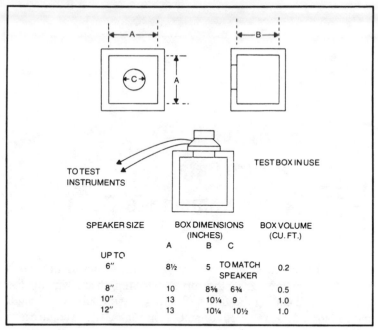

TO TEST
INSTRUMENTS

TEST BOX IN USE

SPEAKER SIZE	BOX DIMENSIONS (INCHES)			BOX VOLUME (CU. FT.)
	A	B	C	
UP TO 6"	8½	5	TO MATCH SPEAKER	0.2
8"	10	8⅝	6¾	0.5
10"	13	10¼	9	1.0
12"	13	10¼	10½	1.0

Fig. B-2. Standard box dimensions for testing speakers of various sizes.

Then total Q (Q_{ts}):

$$Q = \frac{Q_{es} \times Q_{ms}}{Q_{es} + Q_{ms}}$$

Alternate quick method for Q:
Measure R_e, f_s, and Z_{max}.
Find f_1 and f_2 where $Z' = 0.707\, Z_{max}$.
Then:

$$Q = \frac{f_s}{f_2 - f_1} \times \frac{R_e}{Z_{max}}$$

V_{AS}

Hold the speaker over the cut-out in the standard box. (For dimensions of standard box, see Fig. B-2.) Apply enough pressure to make a good seal. Measure the frequency of resonance on the box. Record this frequency as f_{ct}. Find V_{AS} by:

$$V_{AS} = 1.15 \times \left[\left(\frac{f_{ct}}{f_s}\right)^2 - 1 \right] \times V_B$$

171

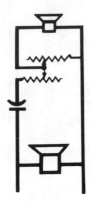

Appendix C
Wire Data
for Homemade Coils

Tests with homemade coils indicate that more consistent results are obtained by measuring the correct wire length and winding the coil without counting the turns. For this reason only wire length and coil form data are given (Figs. C-1 through C-4). In some cases approximations were made to make chart interpretation easier, but tests with various coils made from these charts showed good accuracy.

If you don't have the gauge of wire specified for a coil, you can substitute a wire of larger diameter, or lower gauge number. Only even gauge numbers are listed in these charts, and if you substitute the next lower even number, say #20 for #22, add 10% to the specified length. Avoid using wire of smaller diameter than that listed for a choke, if possible.

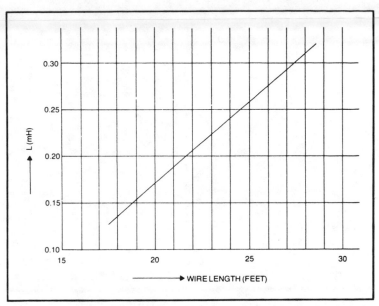

Fig. C-1. Wire length for homemade coils with inductances from 0.1 to 0.32 mH. Use #24 wire on 1/2″ core.

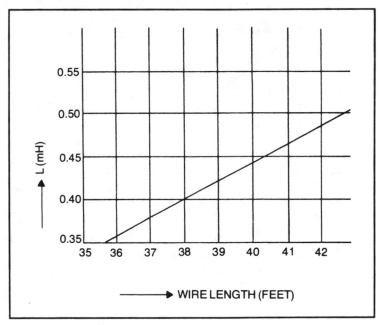

Fig. C-2. Wire length for homemade coils with inductances from 0.33 to 0.51 mH. Use #22 wire on 3/4″ core.

173

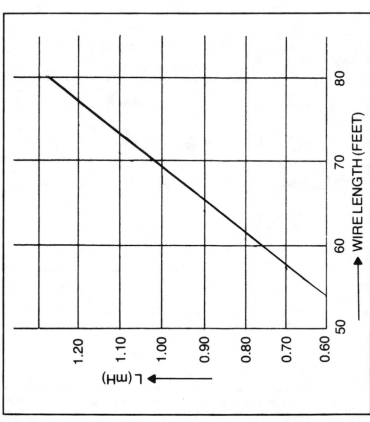

Fig. C-3. Wire length for homemade coils with inductances from 0.60 to 1.28 mH. Use #20 wire on a 1″ core.

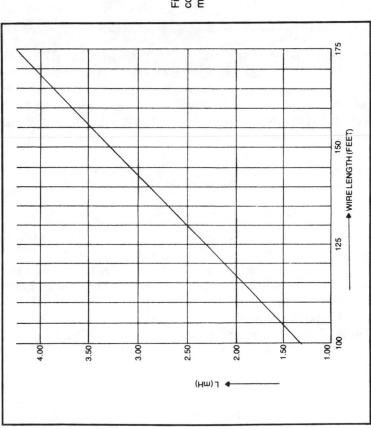

Fig. C-4. Wire length for homemade coils with inductances from 1.4 to 4.3 mH. Use #18 wire on a 1-1/2" core.

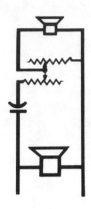

Appendix D
Useful Speaker Design
Formulas and Procedures

Here are two methods of obtaining more detailed information about your vented box design than is available from the simplified charts in Chapter 4. The first is a pocket calculator method of designing a vented box, developed by D. B. Keele, Jr. (Figs. D-1 and D-2.) The second is a computer design program for the Radio Shack TRS-80, written by Robert L. Caudle and based on the equations of Keele and Richard Small (Fig. D-3) As you will see, these procedures permit you to explore how changes in box design can be expected to affect performance.

In order to apply these procedures, you must know the values of f_s, Q, and V_{AS} for your speakers. Table D-1 shows those values for some current Radio Shack woofers.

KEELE'S POCKET CALCULATOR METHOD OF VENTED BOX DESIGN

It takes two flow charts to explain the idea behind this method of vented box design. Fig. D-1 shows the general procedure; Fig. D-2 illustrates the catchy part. If no changes need to be made in the box size, you use the first set of equations for f_3 and f_B; if any changes are made, you use the second set.

General Form

Given: Driver Thiele/Small Parameters
    1. f_s = the frequency of free air resonance
    2. $\mathring{Q}$ = the total driver Q, *sometimes* labeled "Q_{ts}"
    3. V_{as} = the compliance equivalent volume.
Find:
    1. V_B (ft.³), the net box volume

2. f_B (Hz), box frequency of resonance
3. f_3 (Hz), the system cut-off frequency, down 3 dB
4. Hump (or dip) in passband.

Procedure:

$$V_B = 15\, Q^{2.87} V_{as}$$

If no change in box size:

1. $f_3 = 0.26\, Q^{-1.4}\, f_s$
2. $f_B = 0.42\, Q^{-0.9}\, f_s$
3. The design is optimum. There is no hump or dip.

If any change is made in box size:

$$1.\; f_3 = \left(\sqrt{\frac{V_{as}}{V_B}}\, \right) (f_s)$$

$$2.\; f_B = \left(\frac{V_{as}}{V_B} \right)^{0.32} (f_s)$$

$$3.\; \text{Hump (dB)} = 20 \log \left[\, 2.6\, Q \left(\frac{V_{as}}{V_B} \right)^{0.35} \right]$$

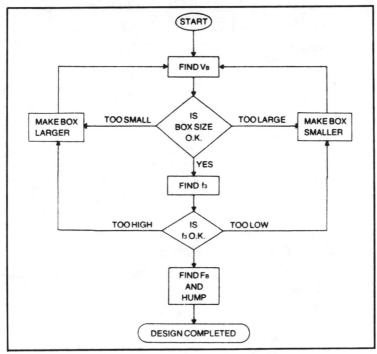

Fig. D-1. Oversimplified flowchart of Keele's pocket calculator method of vented box design.

177

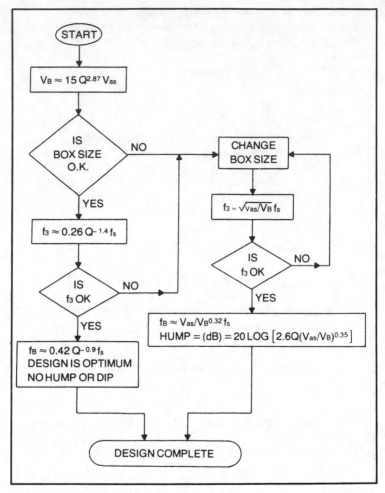

Fig. D-2. Simplified flowchart of Keele's pocket calculator method of vented box design.

KEELE'S CALCULATOR EQUATION
FOR VENTED BOX FREQUENCY RESPONSE

This is another useful mathematical model that goes quickly with a hand calculator.

General Form

Given:

1. Driver parameters

a. f_s (Hz) = free air resonance

b. Q = total driver Q, sometimes labeled "Q_{ts}"

Table D-1. Parameters of Some Radio Shack Speakers.

Cat. No.	Size (in.)	fs (Hz)	Q	Vas (Cu. Ft.)
40-1022	4	55	0.35	0.23
40-1009	6½"	57*	0.6*	0.7*
40-1006	8	43	0.46	2.15
40-1021	8	40	0.41	2.55
40-1331B	10	40	0.93	4.3
40-1023	12	20	0.87	10.8

* These values were obtained by actual test. They vary from those supplied with the speaker.

      c. V_{as} (ft.^{3}) = compliance equivalent volume

2. Box parameters

      a. V_B, net box volume

      b. f_B, box frequency of resonance.

Compute:

1. Constants

      a. $A = f_B^2/f_s^2$

      b. $B = A/Q + f_B/(7f_s)$

      c. $C = 1 + A + f_B/(7f_s Q) + V_{as}/V_B$

      d. $D = 1/Q + f_B/(7f_s)$

2. Response in dB (Normalized to 0 dB at high frequencies)

      a. At each frequency f(Hz) Find: $f_n = f/f_s$

      b. Substitute into:

Response dB = 20 Log.

$$\frac{f_n^4}{\sqrt{(f_n^4 - Cf_n^2 + A)^2 + (Bf_n - Df_n^3)^2}}$$

CAUDLE'S LOUDSPEAKER COMPUTER DESIGN PROGRAM FOR RADIO SHACK TRS-80

The following computer program was written by Robert L. Caudle. Based on the equations of Keele and Richard Small, it is especially designed for the Radio Shack TRS-80 microcomputer. See Fig. D-3.

```
10 REM --------- LOUDSPEAKER DESIGN PROGRAMS ---------
15 REM --------- WRITTEN BY ROBERT L. CAUDLE ---------
20 REM --------- BASED ON THE EQUATIONS OF ---------
25 REM --------- D.B. KEELE AND RICHARD SMALL ---------
30 DIM M(200)
40 CLS:PRINT TAB(15)"LOUDSPEAKER SYSTEM DESIGN PROGRAMS":PRINT:PRINT
50 PRINT "1. DRIVER PARAMETERS"
60 PRINT"2. VENTED BOX DESIGN"
70 PRINT"3. CLOSED BOX DESIGN"
80 PRINT"4. END"
90 PRINT:INPUT"SELECTION (1-4) >";P
100 IF P=1 THEN 140
110 IF P=2 THEN 530
120 IF P=3 THEN 1370
130 IF P=4 THEN END
140 CLS:PRINT TAB(23)"DRIVER PARAMETERS":PRINT:PRINT
150 L$="N":R$="N"
160 INPUT"ENTER DRIVER NAME";D$
170 INPUT"ENTER D.C. RESISTANCE OF VOICE COIL";RE
180 INPUT"ENTER FREE-AIR RESONANCE"; FS
190 INPUT"ENTER IMPEDANCE AT FREE-AIR RESONANCE";ZMAX
200 RO=ZMAX/RE
210 RF=SQR(RO)*RE
220 PRINT"ENTER FREQ. BELOW FREE-AIR =";RF;" OHMS";
230 INPUT F1
240 PRINT"ENTER FREQ. ABOVE FREE-AIR =";RF;" OHMS";
250 INPUT F2
260 QMS=FS*SQR(RO)/(F2-F1)
270 QES=QMS/(RO-1)
```

180

```
280  QTS=QMS*QES/(QMS+QES)
290  INPUT"ENTER TEST BOX VOLUME";TVB
300  INPUT"ENTER DRIVER RESONANCE IN TEST BOX";TFS
310  VAS=TVB*(1.149*((TFS/FS)[2-1))
320  CLS:PRINT TAB(23)"DRIVER PARAMETERS":PRINT:PRINT
330  PRINT D$
340  PRINT"VOICE COIL RESISTANCE(RE)=";RE;" OHMS"
350  PRINT"FREE-AIR RESONANCE(FS)=";FS;" HZ"
360  PRINT"QMS=";QMS
370  PRINT"QES=";QES
380  PRINT"QTS=";QTS
390  PRINT"VAS=";VAS
400  PRINT:PRINT:INPUT"LINE PRINT OUTPUT (Y) ";L$
410  IF L$ <> "Y" THEN 500
420  LPRINT TAB(23)"DRIVER PARAMETERS"
430  LPRINT:LPRINT:LPRINT:LPRINT D$
440  LPRINT"VOICE COIL RESISTANCE (RE)=";RE;" OHMS"
450  LPRINT"FREE-AIR RESONANCE (FS)=";FS;" HZ"
460  LPRINT"QMS=";QMS
470  LPRINT"QES=";QES
480  LPRINT"QTS=";QTS
490  LPRINT"VAS=";VAS
500  INPUT"ANOTHER DRIVER (Y)";R$
510  IF R$="Y" THEN 140
520  GOTO 40
530  CLS:PRINT TAB(23)"VENTED BOX DESIGN":PRINT:PRINT
540  L$="N"
550  INPUT"DRIVER NAME ";D$
560  INPUT"ENTER QTS";QTS
```

Fig. D-3. Robert L. Caudle's loudspeaker computer program for the TRS-80.

181

```
570 INPUT"ENTER VAS";VAS
580 INPUT"ENTER FS";FS

590 VB=15*QTS[2.87*VAS
600 FB=.42*QTS[-.9*FS
610 FH=.26*QTS[-1.4*FS
620 CLS : R$="N"
630 PRINT TAB(25)"B4-ALIGNMENT":PRINT:PRINT
640 PRINT"VB=";VB
650 PRINT"FB=";FB;" HZ"
660 PRINT"F3=";FH;" HZ"
670 H=0
680 PRINT:PRINT:INPUT"CHANGE BOX SIZE (Y)";R$
690 IF R$ <> "Y" THEN 740
700 PRINT:INPUT"ENTER NEW VB ";VB
710 FB=FS*(VAS/VB)[.32
720 FH=FS*SQR(VAS/VB)
730 H=20*LOG(2.6*QTS*(VAS/VB)[.35)/LOG(10)
740 CLS:PRINT TAB(23)"VENTED BOX DESIGN":PRINT:PRINT
750 B$="VENT" : R$="N":L$="N"
760 PRINT D$
770 PRINT"VB=";VB
780 PRINT"FB=";FB;" HZ"
790 PRINT"F3=";FH;" HZ"
800 PRINT"PEAK OR DIP IN RESPONSE =";H;" DB"
810 PRINT:INPUT"CHANGE BOX SIZE (Y)";R$
820 IF R$="Y" THEN 700
830 INPUT"LINE PRINT OUTPUT (Y)";L$
840 IF L$ <> "Y" THEN 900
850 LPRINT TAB(23)"VENTED BOX DESIGN":LPRINT:LPRINT
```

```
855  LPRINT D$
860  LPRINT"VB=";VB
870  LPRINT"FB=";FB;"HZ"
880  LPRINT"F3=";FH;"HZ"
890  LPRINT"PEAK OF DIP IN RESPONSE=";H;"DB"
900  INPUT"VENTED BOX RESPONSE GRAPH (Y)";R$
910  IF R$="Y" THEN 1020
920  INPUT"CUSTOM DESIGN (X)";R$
930  IF R$ <> "Y" THEN 1340
940  CLS:PRINT TAB(20)"CUSTOM VENTED BOX DESIGN"
950  B$="" : R$="N"
960  PRINT:PRINT:INPUT"CHANGE VB (Y)";R$
970  IF R$<> "Y" THEN 985
980  INPUT"ENTER NEW VB ";VB
985  R$="N"
990  INPUT"CHANGE BOX TUNING (Y)";R$
1000 IF R$<> "Y" THEN 1020
1010 INPUT"ENTER NEW FB ";FB
1020 CLS:PRINT TAB(15)"GRAPHICS CALCULATION IN PROGRESS"
1030 A=(FB[2)/(FS[2)
1040 B=A/QTS+(FB/(7*FS))
1050 C=1+A+(FB/(7*FS*QTS))+(VAS/VB)
1060 D=1/QTS+(FB/(7*FS))
1070 FOR F=20 TO 200 STEP 5
1080 F9=F/FS:F5=F9[2
1090 F4=F9[4:F3=F9[3
1100 F6=(F4-C*F5+A)[2
1110 F7=(B*F9-D*F3)[2
1120 M(F)=20*(LOG(F4/(F6+F7)[.5)/LOG(10))
1130 NEXT
1140 GOSUB 1900
```

Fig. D-3. Robert L. Caudle's loudspeaker computer program for the TRS-80 (continued from page 181).

183

```
1150 IF B$="VENT" THEN 740
1160 CLS:PRINT TAB(20)"CUSTOM VENTED BOX DESIGN"
1165 R$="N"
1170 PRINT:PRINT:PRINT D$
1180 PRINT"VB=";VB
1190 PRINT"FB=";FB;" HZ"
1200 PRINT"QTS=";;QTS
1210 PRINT"FS=";FS;" HZ"
1220 PRINT"VAS=";VAS
1230 PRINT:PRINT:INPUT"CHANGE VB OR FB (Y)";R$
1240 IF R$="Y" THEN 940
1250 INPUT"LINE PRINT OUTPUT (Y)";L$
1260 IF L$ <> "Y" THEN 1340
1270 LPRINT TAB(20)"CUSTOM VENTED BOX DESIGN":LPRINT:LPRINT
1280 LPRINT D$
1290 LPRINT"VB=";VB
1300 LPRINT"FB=";FB;"HZ"
1310 LPRINT"QTS=";;QTS
1320 LPRINT"FS=";FS;"HZ"
1330 LPRINT"VAS=";;VAS
1340 INPUT"ANOTHER VENTED BOX DESIGN (Y)";R$
1350 IF R$="Y" THEN 530
1360 GOTO 40
1370 CLS:PRINT TAB(23)"CLOSED BOX DESIGN":PRINT:PRINT
1380 L$="N" : R$="N"
1390 INPUT"DRIVER NAME";D$
1400 INPUT"ENTER QTS";QS
1410 INPUT"ENTER VAS";VAS
1420 INPUT"ENTER FS";FS
1430 INPUT"ENTER VB";VB
1440 A=VAS/VB
```

```
1450 FC=FS*SQR(A+1)
1460 QTC=(FC*QS)/FS
1470 F3=FC*SQR(((1/QTC[2-2)+SQR((1/QTC[2-2)[2+4))/2)
1480 CLS:PRINT TAB(23)"CLOSED BOX DESIGN"
1485 R$="N"
1490 PRINT:PRINT:PRINT D$
1500 PRINT"F3= ";F3;" HZ"
1510 PRINT"QTC= ";QTC
1520 PRINT"VB= ";VB
1530 PRINT:INPUT"CHANGE BOX SIZE (Y)";R$
1540 IF R$ <> "Y" THEN 1570
1550 PRINT:INPUT"ENTER NEW VB ";VB
1560 GOTO 1440
1570 CLS:PRINT TAB(23)"CLOSED BOX DESIGN"
1580 R$="N":L$="N"
1590 PRINT:PRINT:PRINT D$
1600 PRINT"QTS=";QS
1610 PRINT"QTC=";QTC
1620 PRINT"VAS=";VAS
1630 PRINT"VB=";VB
1640 PRINT"FS=";FS;" HZ"
1650 PRINT"ALPHA=";A
1660 PRINT"FC=";FC;" HZ"
1670 PRINT"F3=";F3;"HZ"
1680 PRINT:INPUT"LINE PRINT OUTPUT (Y)" ;L$
1690 IF L$ <> "Y" THEN 1780
1700 LPRINT TAB(22)"CLOSED BOX DESIGN":LPRINT:LPRINT
1710 LPRINT"QTS=";QS
1720 LPRINT"QTC=";QTC
```

Fig. D-3. Robert L. Caudle's loudspeaker computer program for the TRS-80 (continued from page 183).

```
1730 LPRINT"VAS=";VAS
1740 LPRINT"VB =";VB
1750 LPRINT"FS=";FS;"HZ"
1760 LPRINT"ALPHA=";A
1770 LPRINT"FC=";FC;"HZ"
1775 LPRINT"F3=";F3;"HZ"
1780 PRINT:INPUT"CLOSED BOX RESPONSE GRAPH (Y)";R$
1790 IF R$ <> "Y" THEN 1865
1800 CLS:PRINT TAB(15)"GRAPHICS CALCULATION IN PROGRESS"
1810 FOR F=20 TO 200 STEP 5
1820 FH=F/FC:FQ=FH[2:MAG=FQ/(SQR((FQ-1)[2+(FH/QTC)[2))
1830 M(F)=20*(LOG(MAG)/LOG(10))
1840 NEXT
1850 GOSUB 1900
1860 GOTO 1570
1865 R$="N"
1870 INPUT"ANOTHER CLOSED BOX DESIGN (Y)";R$
1880 IF R$="Y" THEN 1370
1890 GOTO 40
1900 CLS:RESTORE:R$="N"
1910 PRINT:PRINT:PRINT:PRINT"0 DB":PRINT:PRINT:PRINT"5 DB":PRINT:PRINT"PER": PRINT"DIV"
1920 PRINT:PRINT:PRINT
1930 PRINT TAB(5)"20" TAB(16)"HZ" TAB(26)"200"
1940 FOR I=12 TO 54
1950 SET(I,38)
1960 NEXT
1970 FOR I=3 TO 38
1980 SET(12,I)
```

186

```
1990 NEXT
2000 FOR I=3 TO 33 STEP 5
2010 SET(13,I)
2020 NEXT
2030 FOR I=1 TO 9
2040 READ J
2050 SET(LOG(J/20)*18.3851)+12,37)
2060 NEXT
2070 FOR F=20 TO 200 STEP 5
2080 IF M(F) <-25 THEN M(F) = -25
2090 SET(LOG(F/20)*18.3851)+12,(38-M(F))-25)
2100 NEXT
2110 DATA 30,40,50,60,70,80,90,100,200
2115 DATA 20,30,40,45,50,60,80,100,120,140,160,200
2120 INPUT"LINE PRINT (Y) ";R$
2130 IF R$="Y" THEN 2150
2140 RETURN
2150 LPRINT TAB(9)"-40" TAB(19)"-30" TAB(29)"-20" TAB(39)"-10" TAB(50)"0" TAB(59)"+10"
2160 LPRINT TAB(10)"+" TAB(20)"+" TAB(30)"+" TAB(40)"+" TAB(50)"+" TAB(60)"+"
2170 FOR I=1 TO 12
2180 READ F
2190 LPRINT F TAB(9)"I";
2200 IF M(F)<-39) THEN .2230
2210 LPRINT TAB(50+M(F))"*"
2220 GOTO 2240
2230 LPRINT ""
2240 NEXT
2250 INPUT"PRESS ENTER TO CONTINUE"; R$
2260 RETURN
```

Fig. D-3. Robert L. Caulle's loudspeaker computer program for the TRS-80 (continued from page 185).

Index

189